WINE TRAILS
EUROPE

PLAN 40 PERFECT WEEKENDS IN WINE COUNTRY

INTRODUCTION

We've all experienced it on our travels - whether watching a game of pétanque in Provence with a chilled rosé, or at a tapas bar in Andalucía with a dry and savoury Fino sherry - when a local wine could not be more perfectly suited to the moment.

Tasting wine in the place it was made can be a revelation. This book plots a course through 40 of Europe's greatest wine regions, with weekend-long itineraries in each. We encounter intense red wines in Friuli and crisp and fruity Rieslings produced on the banks of the Mosel and Rhine rivers. We venture into historic, world-famous wineries and cutting-edge cellar doors, and in Portugal's Alentejo and Georgia's Kakheti region we discover some unsung heroes. In each region, our expert writers - including Masters of Wine Caroline Gilby and Anne Krebiehl and critics and columnists Sarah Ahmed, Tara Q. Thomas and John Brunton - review the most rewarding wineries to visit and the most memorable wines to taste.

This is a book for casual quaffers; there's no impenetrable language about malolactic fermentation or scoring systems. Instead, we meet some of the world's most enthusiastic and knowledgeable winemakers and learn about each region's wines in their own words. It is this personal introduction to wine, in its home, that is at the heart of wine-touring's appeal.

CONTENTS

EUROPE
NORTH SEA
SWEDEN
DENMARK
COPENHAGEN
IRELAND
ENGLAND
WALES
CARDIFF
LONDON
SOUTHAMPTON
04
ENGLISH CHANNEL
AMSTERDAM
NETHERLANDS
BRUSSELS
BELGIUM
02
16
KOBLENZ
LUXEM-
BOURG
15
GERMANY
BERLIN
17
LEIPZIG
PRAGUE
CZECH REPUBLIC
REIMS
PARIS
08
STRASBOURG
05
VIENNA
01
AUSTRIA
ORLEANS
11
DIJON
07
BERN
SWITZERLAND
LIECH-
TENSTEIN
21
FRANCE
09
THE ALPS
TRENTO
22
SLOVENIA
LJUBLJANA
TRIESTE
33
03
NORTH
ATLANTIC
OCEAN
LYON
13
TURIN
24
29
BAY OF
BISCAY
06
BORDEAUX
12
23
GENOA
SAN
MARINO
TOULOUSE
10
MONACO
FLORENCE
MARSEILLE
SANTIAGO DE
COMPOSTELA
34
40
38
BURGOS
ANDORRA
28
ITALY
ADRIATIC
SEA
Corsica
(FRANCE)
ROME
PORTO
31
39
37
BARCELONA
26
MADRID
BALEARIC
SEA
PORTUGAL
SPAIN
PALMA
36
Sardinia
(ITALY)
TYRRHENIAN
SEA
30
CAGLIARI
LISBON
EVORA
MEDITERRANEAN SEA
27
Sicily
(ITALY)
CATANIA
SEVILLE
35
Gibraltar (UK)
ALGIERS
TUNIS
MALTA
TUNISIA
ALGERIA
RABAT
MOROCCO

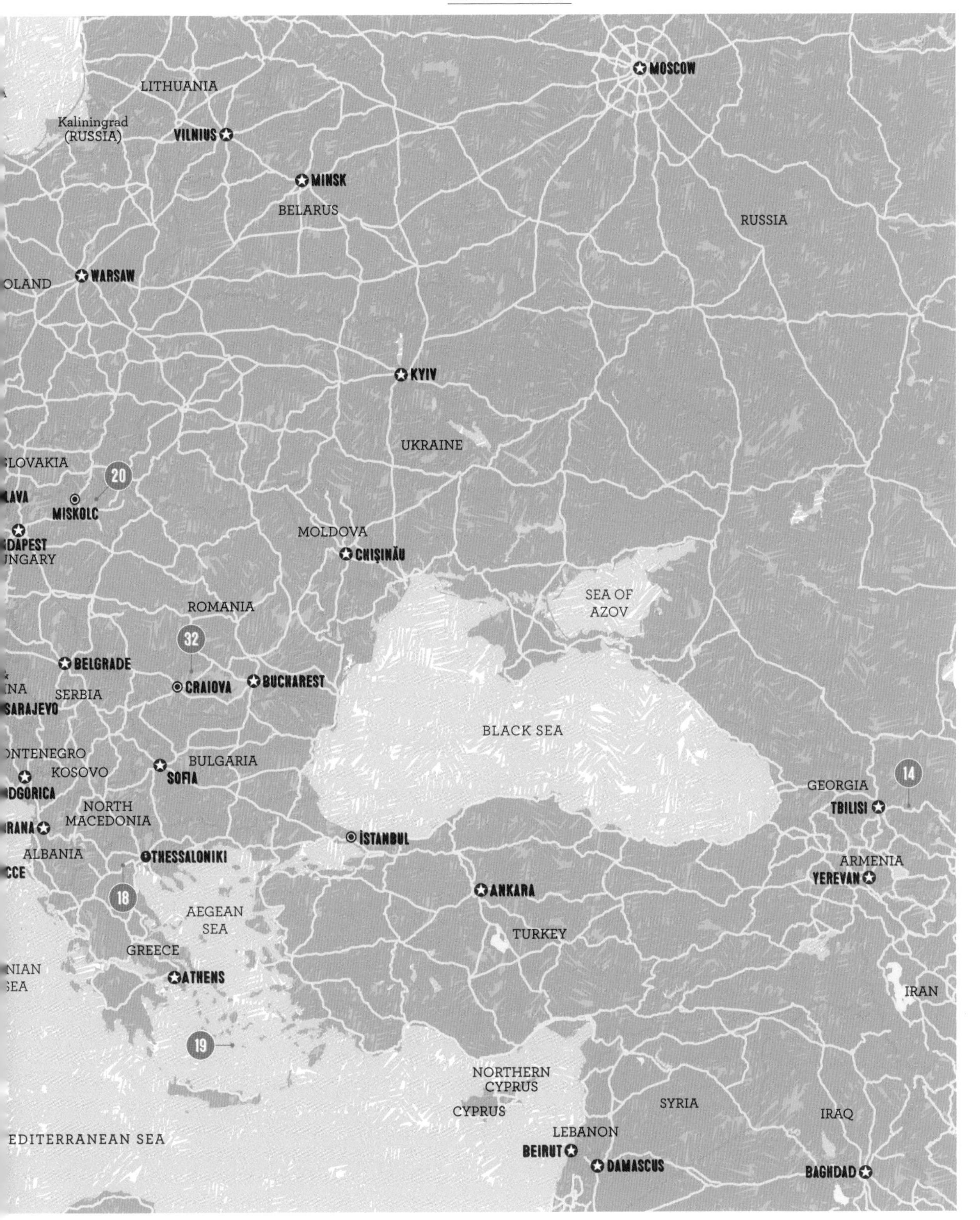
MOSCOW
LITHUANIA
Kaliningrad (RUSSIA)
VILNIUS
MINSK
BELARUS
RUSSIA
WARSAW
OLAND
KYIV
UKRAINE
LOVAKIA
20
AVA
MISKOLC
DAPEST
UNGARY
MOLDOVA
CHIŞINĂU
SEA OF AZOV
ROMANIA
32
BELGRADE
CRAIOVA
BUCHAREST
NA
SERBIA
SARAJEVO
BLACK SEA
NTENEGRO
KOSOVO
SOFIA
BULGARIA
DGORICA
14
GEORGIA
TBILISI
NORTH MACEDONIA
RANA
İSTANBUL
ALBANIA
THESSALONIKI
ARMENIA
YEREVAN
CCE
18
ANKARA
AEGEAN SEA
TURKEY
GREECE
NIAN SEA
ATHENS
IRAN
19
NORTHERN CYPRUS
SYRIA
CYPRUS
IRAQ
EDITERRANEAN SEA
LEBANON
BEIRUT
DAMASCUS
BAGHDAD

01

[Austria]

VIENNA

As the world's only capital growing significant amounts of wine within its city limits, Vienna is a one-off with its urban-edge vines and historic Heuriger.

There's a certain thrill about hopping on the tram or U-Bahn, leaving central Vienna's hustle-bustle behind, and within half an hour striding among vineyards that march down the hillsides to the ribboning Danube. The wineries sprinkled across Vienna's 18th, 19th, 21st and 23rd districts make a smooth transition between the culture-loaded capital and the great outdoors, with most fringing the undulating woodlands of the Wienerwald Unesco Biosphere Reserve.

Many of Vienna's *Weingüter* (wineries and estates) can trace their origins back to the 17th century or beyond, and are still family-run today. It's not unusual to get chatting with a vintner and hear tales of how Hapsburg royals praised their Rieslings, how Mozart quaffed wine in their *Heuriger* (traditional Viennese wine tavern), or how Beethoven found inspiration for his greatest symphonies right here in these vineyards.

Open when you see the *Ausg'steckt* sign and a bunch of fir branches, the *Heuriger* is an integral part of the wine experience in Vienna. On hazy summer evenings, when the city appears like a smudged pencil drawing on the horizon and the Danube glints silver below, locals come to indulge in wine and buffet food in snug wood-panelled parlours and vine-draped gardens.

In the here and now, Viennese wine is having a moment. The popularity of its unique, protected origin Gemischter Satz (field blend) wines is booming. And trailblazing vintners are putting their own riffs on biodynamic wines that hit all the right sustainable notes. Around 80% of the wines are white – Riesling, Pinot Blanc, Grüner Veltliner, Gelbe Muskateller and the like – but reds such as Zweigelt, St Laurent and Pinot Noir are going from strength to strength, too.

GET THERE
Vienna airport, 18km southeast of the city centre, has dozens of flights to destinations across Europe.

01 COBENZL

Vienna opens up like a pop-up book before you from the viewpoint at Cobenzl, bordering the Wienerwald woods in the 19th district, where serried rows of vines make a boldly contrasting leap between the capital and the great outdoors. Spread across the hills of 382m (12,532ft) Reisenberg and 492m (1614ft) Latisberg, these vineyards are terrific for a wander.

Walking is thirsty work, *natürlich*, which is where this winery fits into the picture. Thomas Podsednik heads up a slick operation, combining family-honed tradition with ultra-modern equipment and sustainable methods (solar power, compost as fertiliser, etc). The resulting vineyards are among Austria's finest, three meeting the strict criteria for *Erste Lage* (first-class site) status.

Blown-up photos of vines grace the backlit tasting room and shop, where you can opt for a three-wine tasting. Top billing goes to single-vineyard wines like Ried Steinberg Gemischter Satz, a dry, powerful, mineral-packed white, with aromas of apricot and pear, that's perfect for ageing. The 24-month oak barrel-aged Pinot Noir, with aromas of chestnut and red berries, also comes highly recommended. For times and details on guided tours of the press house, historic cellar and vineyards, see the website.

www.weingutcobenzl.at; tel 01 320 58 05; Am Cobenzl 96, 1190 Vienna; Mon-Fri $

02 CHRIST

It might look proudly back on 400 years of family winegrowing tradition but that hasn't stopped Christ from becoming one of Vienna's most progressive wineries. On the vine-clad hill of Bisamberg in the 21st district, Christ manages the delicate act of bringing past and present together with a clean, modern aesthetic that prizes such natural materials as wood and stone. The winemakers use, for instance, both oak barrels and stainless-steel wine tanks for fermentation, depending on the results they wish to achieve.

The mineral-rich soil here (a mix of glacial gravel, weathered limestone and slate) plays its part, too, in yielding top-quality

01 Vineyards near Vienna

02 Vine-strewn exterior of Mayer am Pfarrplatz

03 Traditional *Heuriger* sign

04 Cobenzl winery

grapes that go into the likes of an organic sparkling white with fresh citrus notes, a lightly floral, vegan rosé, and the more complex, full-bodied, silky Mephisto (a Merlot, Cabernet Sauvignon and Zweigelt blend), hinting at dark chocolate aromas. Taste and buy them in the contemporary wine store, or in the *Heuriger* paired with dishes such as black pudding carpaccio with horseradish, or spinach and sheep's cheese strudel. The garden shaded by vines and oleander is a peaceful escape from the Vienna city buzz during summer. ***www.weingut-christ.at; tel 01 292 51 52; Amtsstrasse 10–14, 1210 Vienna; daily Jan, Mar, May, Jul, Sep & Nov or by appointment at other times*** $

03 EDLMOSER

This feted winery in Mauer in the 23rd district nudges up against the Lainzer Tiergarten, a former imperial hunting ground turned wildlife reserve. It's quite remarkable to find yourself in such natural surrounds just a half-hour trundle on tram 60 from Vienna's Westbahnhof. With a family winegrowing tradition harking back to 1374, there's plenty of history to the place, as well as sweeping views over vine and city, and an excellent terroir, with Veltliner white. Punchier, however, is the Vienna blend (Zweigelt and Sankt Laurent), a vibrant, elegant, spicy red. Michael Edlmoser is as passionate about music as wine, featuring pairing playlists on the website. Often ranked Vienna's best *Heuriger*, the rustic tavern dates to 1629 and gives onto a rambling, vine-swathed garden. Here the wines are matched with buffet dishes such as caraway roast and wild boar sausage with nut bread. ***www.edlmoser.com; tel 01 889 86 80; Maurer Lange Gasse 123, 1230 Vienna; Thu–Sun***

04 ITZ WIENINGER

The ancient brick-vaulted cellars at this winery in Stammersdorf in the 21st district once belonged to a monastery. Now the monks are long gone and in their place is forward-thinking vintner Fritz Wieninger, a champion of Vienna's one-of-a-kind Gemischter Satz wines and a firm believer

in biodynamic production and organic farming methods.

The light, sandy, windblown silt soil up here is ideal for Burgundy-style varietals. Perhaps star of the show is the highly aromatic Bisamberg Gemischter Satz, with good minerality and an intensely spicy nose (try it with dark bread, cheese and charcuterie). Also worth a mention are the likes of the bright, juicy Pinot Noir with plum and cherry aromas, and the fresh, peachy Gelber Muskateller. There's no finer place to try them than with a mixed platter of pork roast, mountain cheese, local hams, pickles and dark bread in the cosy wood-panelled *Heuriger*. Lanterns cast a romantic glow over the vine-canopied garden after dark.

http://heuriger-wieninger.at; tel 01 292 41 06; Stammersdorfer Strasse 78, 1210 Vienna; Fri–Sun late Mar–mid Dec

05 MAYER AM PFARRPLATZ

Spread across a cluster of historic buildings – one of which Beethoven called home in 1817 when he was busy composing his magnum opus Ninth Symphony – in Heiligenstadt in the 19th district, Mayer am Pfarrplatz has been going strong since 1683. It prides itself today on its single sites, chalky soils and old vines, as well as hand-picking and fermentation in stainless-steel tanks, all of which define the characteristics of its sought-after, fruit-driven Gemischter Satz, Riesling and Grüner Veltliner wines.

With its rustic interior and garden shaded by vines and walnut trees, the *Heuriger* is hands-down one of Vienna's loveliest. Here you can sample the wines while digging into seasonally inspired dishes and classics such as organic pork roast with sauerkraut and sweet apricot dumplings. Live music performances happen regularly (including jazz and traditional Austrian) and there's a playground to keep kids entertained.

www.pfarrplatz.at; tel 01 370 12 87; Pfarrplatz 2, 1190 Vienna; Mon–Sun

05 Alfresco *Heuriger*

06 Viennese wine tavern

ESSENTIAL INFORMATION

WHERE TO STAY

HOTEL RATHAUS WEIN & DESIGN

The clue's in the name: this boutique hotel in Vienna's happening 8th district is all about wine. Each minimalist room is named after an Austrian winery, toiletries are grape infused, and the minibar features wines from your vintner host. Best of all is the wine bar, which spotlights a different Austrian winery every month. *www.hotel-rathaus-wien.at; tel 01 400 11 22; Lange Gasse 13, 1080 Vienna*

HOTEL TOPAZZ

With a monochrome palette, geometric motifs and oval-shaped windows, Hotel Topazz does understated chic with a pinch of glamour. It's owned by the Lenikus group, whose excellent organic wines you can enjoy over knockout views of Vienna at the rooftop bar of sister hotel Lamée, opposite. *www.hoteltopazz.com; tel 01 532 22 50; Lichtensteg 3, 1010 Vienna*

06

WHERE TO EAT

GASTHAUS GRÜNAUER

A good old-fashioned, family-run Viennese tavern, with a vaulted, wood-lined interior and booth seating, Grünauer dishes up spot-on Austrian classics such as schnitzel and goulash, marrying them with an excellent selection of some 200 Austrian wines. *www.gasthaus-gruenauer.com; tel 01 526 40 80; Hermanngasse 32, 1070 Vienna*

PFARRWIRT

Lodged in a beautiful 12th-century building surrounded by vineyards, Pfarrwirt puts a creative touch on traditional cuisine in such dishes as Danube beef carpaccio with lime cream and chanterelles. It also rustles up a tasty *Tafelspitz* (boiled beef with apple-horseradish sauce). There's a strong focus on Viennese wines. *www.pfarrwirt.com; tel 01 370 73 73; Pfarrplatz 5, 1190 Vienna*

WHAT TO DO

SCHLOSS SCHÖNBRUNN

The Unesco-listed former summer palace of the Hapsburgs is a lavish, 1441-room reflection of their refined tastes. In front of the baroque orangerie (the world's second largest after Versailles), you'll find a lovingly recreated vineyard, which produces a fine line of Gemischter Satz wines sold at auction. *www.schoenbrunn.at*

VILLON

Right in the heart of the 1st district, this 500-year-old wine cellar is sunk 16m (52ft) deep into the ground. Order wine by the glass or bottle (along with tasting platters), or join a guided tour or wine tasting, with 44 open wines to sample. *www.villon.at*

CELEBRATIONS

VIENNA WINE HIKING DAY

With the smell of must in the air, the harvest underway and the landscape an embroidery of gold and russet, Vienna's vineyards are a delight in autumn. On a late-September day, this event invites you to stride on three routes totalling 25km (16 miles), with vintners offering tastings en route.

VIENNA WINE FEST

Wine is culture at this May festival held in the city's colossal MuseumsQuartier: after an art fix, meet producers and taste wines from more than 60 vineyards. *www.viennawinefest.com*

01

BELGIUM & LUXEMBOURG

Seek out classics and avant-garde clones among the tiny, hidden vineyards of Belgium and the picturesque riverside terraces of Luxembourg.

Beer and chocolate may be the first things that spring to mind, but Belgium has a strong claim to the title of world's smallest wine-producing country, with 200 to 300 hectares (500 to 750 acres) under cultivation. A visit to this cold-climate region promises encounters with enthusiastic, eccentric *vignerons*, well-organised wine routes, a long but little-known history of viticulture, and a host of quality white, red and sparkling vintages.

The most interesting producers, independent artisan *vignerons*, are concentrated in French-speaking Wallonia, around the cities of Mons, Liège and Namur, where vineyards were first grown as far back as the 9th century. Most disappeared during the Middle Ages when temperatures cooled and brewing beer became popular. Only now – with global warming, advances in vinification technology and cloned 'interspecific' grapes that are disease-resistant – is Belgium back on the world wine map. Travelling through the bucolic countryside you'll find evidence of two contrasting wine philosophies: classicists growing Chardonnay and Pinot Noir who are inspired by French Champagne and Burgundy; and modernists planting the new generation of clones such as Solaris, Regent and Johanniter.

In tiny Luxembourg, a 42km (26-mile) stretch along the steep terraced banks of the Moselle has been producing grapes continuously since Roman times. With over 1200 hectares (2965 acres) under production, it is only in the last twenty to thirty years that some 55 independent *vignerons* have stopped selling grapes to the all-powerful Caves Coopératives to focus on bottling their own cuvées. Along the picturesque Moselle wine route, you can taste excellent sparkling Crémant, but also crisp, fruity Riesling, Pinot Gris and aromatic Müller-Thurgau, plus Pinot Noir well worthy of the *vignobles* of Alsace.

GET THERE
Brussels-Charleroi Airport is 40km (25 miles) from Haulchin. Car hire is available.

© Pecold | Shutterstock

01 DOMAINE DES AGAISES

The vineyards of Domaine des Agaises bear more than a passing resemblance to France's Champagne region. Those renowned Champagne domaines are just 90km (56 miles) away, with the same climate, the same chalky soil. The bunches of plump grapes growing at Domaine des Agaises just happen to be Chardonnay, Pinot Noir and Meunier – the magic combination behind the world's most famous sparkling wine. Winemaker Raymond Leroy and his sons, however, are putting them to work to create Belgium's very own bubbly. The label on the distinctive Ruffus bottle says *Méthode Traditionelle* (the *Méthode Champenoise* description is fiercely protected by France) but when the cork is popped and the bubbles rise up in a crystal flute, surely only experts could tell the difference.

www.ruffus.be; tel 0497 88 53 10; 1 Chemin d'Harmignies, Haulchin, Belgium; Mon–Fri by appointment $

02 DOMAINE DU CHENOY

A winding vineyard path leads up to the rambling 18th-century farmhouse where Philippe Grafe has created the most dynamic organic winery in Belgium. He pioneered the use of 'interspecific' grapes – hardy clones that require very little chemical treatment – which has inspired a host of young *vignerons* to follow suit. 'Believe it or not,' he recounts, 'the moment when I decided to plant a vineyard was during a trip to Cornwall, where I tasted wine made from the Solaris grape – it was absolutely wonderful and I was sure I could have the same results here in Belgium. I discovered that Solaris was one of these new revolutionary grapes, perfect for our climate and soil conditions.' He is justifiably proud of cuvées like Butte aux Lièvres and Taille aux Renards, made with Regent and Pinotin interspecific grapes, which achieve exceptional quality when left to age for several years.

www.domaine-du-chenoy.com; tel 081 74 67 42; 1B Rue du Chenoy, Emines, Belgium; Mon–Sat by appointment

01 Mons

02 Château de Bioul

03 Tasting at Maison Schmit-Fohl

04 Grape harvest at Domaine des Agaises

'During a trip to Cornwall I tasted a wonderful wine made from the Solaris grape. I was sure I could have the same results in Belgium'

– Philippe Grafe, Domaine du Chenoy

03 CHÂTEAU DE BIOUL

The tiny village of Bioul is dominated by its grandiose castle, family home of Vanessa Vaxelaire, who moved back here from Brussels eight years ago, determined to become a *vigneronne*. She sold off part of the family's farming land to buy 10 hectares (28 acres) favourable for growing grapes. Although not officially organic, the domaine respects nature, with beehives kept in between some vines, scarecrows rather than chemicals to discourage birds, and the choice of grapes like Pinotin and Solaris, which don't need treatments to protect them from mildew. The château's tasting room is festooned with dozens of boars' heads and antlers, but the real excitement comes when you walk into the cellar where, together with her young French wine consultant, Mélanie Chereau, Vanessa uses state-of-the-art technology, such as egg-shaped cement tanks, to create her wine.

www.chateaudebioul.be; tel 071 79 99 43; 1 Place Vaxelaire, Bioul, Belgium; Mon–Fri by appointment & Sat–Sun

04 LA CLOSERIE DES PRÉBENDIERS

La Closerie des Prébendiers looks more like a comfortable retirement villa than a wine domaine, but the moment Jacques Mouton shows visitors his back-garden vineyard, they fall under the charming spell of this passionate *vigneron*. Some 35 years ago, he planted a series

05

05 The Moselle at Schengen

06 Grapes on the vine, Domaine Kox

07 Maison Schmit-Fohl

of terraced vineyards that tumble down almost vertically towards the Meuse river, quite as spectacular as the famed Côte Rôtie in France's Rhône Valley. Vertigo sufferers take note: during the harvest, it is too dangerous to carry the grapes up to the cellar, so instead they are transported by mini rail track using a hydraulic trolley. The domaine produces roughly one thousand bottles of a single Vin Belge, a crisp, fruity white assemblage of Müller-Thurgau, Auxerrois and Pinot Gris grapes.

Tel 085 21 12 23; Thier des Malades 16, Huy, Belgium; daily by appointment $

05 MAISON SCHMIT-FOHL

The picturesque winemaking village of Ahn is the jewel in the crown of Luxembourg's Moselle vineyards, with steep vine-clad stone terraces climbing up from the river's edge. Among the village's one hundred inhabitants there are seven independent *vigneron* families, and 27-year-old Nicolas Schmit-Fohl is the 11th generation to work 33 tiny parcels of vines dotted along the valley. Over 50% of production is white, not the classic Moselle Riesling or Pinot Gris, but a structured, elegant, oak-aged Chardonnay. The bubbly Crémant is also aged in small barrels, 'so the result is much closer to a Champagne,' says Nicolas, adding that, 'more and more young people in Luxembourg are getting interested in winemaking, and here in Ahn, every domaine has been enthusiastically taken over by the new young generation like myself.'

www.schmit-fohl.lu; tel 76 0231; 8 Rue de Niederdonven, Ahn, Luxembourg; daily

06 DOMAINE KOX

In the hills above the Moselle, the Kox family manage the most experimental winery in all of Luxembourg, producing an impressive selection of very surprising wines: a dozen different sparkling Crémants, including zero-dosage and non-sulphite cuvées; a remarkable Pinot Noir that could easily blind-test alongside a Burgundy; orange wine aged in a terracotta amphora from Georgia; and crisp, fresh, fruity Rhäifrensch. 'This is made from the indigenous Eberling grape', explains Corinne Kox, 'which has been grown here for 1000 years but it is barely cultivated any more. My dad says it was known as the "water of the people", because it was safer to drink than water, which was usually contaminated and unhealthy. It is part of our culture and tradition so we are determined to still produce Eberling. We harvest as late as possible, press with our feet, and manage to keep down its natural acidity so Rhäifrensch is perfect to drink outside on a hot summer's day.' Now, you can sample them all during a guided tour of the vineyard.

www.domainekox.lu; tel 2369 8494; 6A Rue des Prés, Remich, Luxembourg; daily by appointment $

ESSENTIAL INFORMATION

WHERE TO STAY

LES TANNEURS
This maze-like historic hotel, housed in a grand 17th-century mansion in Namur, makes a great base for exploring the Meuse Valley vineyards. It's a gourmet restaurant and casual bistro too. *www.tanneurs.com; tel 081 24 00 24; 13 Rue des Tanneries, Namur, Belgium*

AUBERGE DU CHÂTEAU
Located in between vineyards and the Moselle riverbank, this simple and very reasonably priced country inn also houses an excellent restaurant serving local freshwater fish, such as pikeperch in a Riesling sauce or a bowl of *friture*, crunchy deep-fried whitebait. *www.auberge-chateau.lu; tel 2666 41 49; 11 Route du Vin, Stadtbredimus, Luxembourg*

WHERE TO EAT

AUX SAVEURS D'ARDENNE
In the midst of Belgium's thick wooded Ardennes, this traditional tavern serves its speciality trout with almonds, plus venison and wild boar in the hunting season. *Tel 0470 69 38 53; 32 Rue Grande, Vencimont, Belgium*

WÄISTUFF A POSSEN
Next door to a wine and folklore museum, this bistro is a favourite with local *vignerons* for both its extensive Luxembourg wine list and tasty dishes, including speciality *kniddelen* dumplings in a rich creamy sauce. *www.waistuffapossen.lu; tel 2455 84 44; 4 Keeseschgässel, Bech-Kleinmacher, Luxembourg*

VINOTECA
Luxembourg City has a vibrant eating-out scene. Vinoteca's host, award-winning sommelier Rodolphe Chevalier, is an expert at grilling huge tomahawk steaks on his courtyard barbecue. *www.barvinoteca.lu; tel 2668 38 43; 6 Rue Wiltheim, Luxembourg City*

WHAT TO DO

LIÈGE
The best nightlife spot in this fun, studenty city is the totally surreal Pot au Lait bar, with psychedelic murals, fluorescent skulls, neon, and sci-fi and religious statues. *www.potaulait.be*

SPIENNES
Descend into a prehistoric flint mine just outside Mons, first excavated by Neolithic tribes and today a Unesco World Heritage Site. *https://whc.unesco.org*

MONDORF DOMAINE THERMAL
A traditional health spa in a beautiful 19th-century park near the Moselle. Day passes cover the 36°C thermal pools, saunas, jacuzzis and hammams. *www.mondorf.lu*

CELEBRATIONS

BINCHE CARNIVAL
Celebrated around Shrove Tuesday and a rival to Rio, this is when the town's menfolk dress up in their surreal Gilles costumes.

EAT IT
A vibrant food-truck festival held in Luxembourg City each April, serving everything from burgers and curries to vegetarian-friendly and gluten-free delicacies, as well as the local speciality: delicious potato pancakes.

[Croatia]

ISTRIA

The green hills of Istria have produced wine for Roman emperors, Venetian nobles and Austrian royalty for centuries, and local families still keep the tradition alive.

The heart-shaped peninsula at the top of Croatia's achingly beautiful Adriatic coastline has long been lauded for its ancient buildings, pebbly beaches and fairy-tale hilltop towns, but its burgeoning reputation as an epicurean heaven is a relatively new development. In recent years, Istrian olive oils have won international acclaim, while the black and white truffles harvested from the region's enchanting forests have been working their magic, too. These pungent fungi enliven a cuisine deeply indebted to the many centuries when Istria was an integral part of La Serenissima, the once-great Venetian Republic.

It's no surprise that visiting gourmets have woken up to the excellence of Istrian wine as well. *Vino* has been produced in Croatia for millennia, and there's quite an array of highly localised indigenous grape varieties grown throughout the country. Istria's main viticultural treasure is Malvazija Istarska, a versatile white grape that lends itself to a range of styles: from lean and crisp to bold and oaked; dry or sweet; and sometimes even sparkling. On the red side of the ledger is Teran, which yields a rustic, richly hued wine that is best drunk young. Imported varietals that prosper in Istria include white Muscat (known locally as Muškat Momjanski, as it's mainly grown around the town of Momjan, near the Slovenian border), as well as Sauvignon Blanc, Chardonnay, Cabernet Sauvignon and Merlot.

GET THERE
Dozens of airlines fly directly to Pula in summer; otherwise, transit via Zagreb. Catamarans from Venice to Poreč, Rovinj and Pula operate seasonally.

Most of Istria's wine producers are small, family-run affairs – and while they're welcoming of visitors, there isn't the same network of cellar-door tasting rooms that you'll find in the more established wine-touring regions. That's part of the charm: driving up to small vineyards where – perhaps with only a little English spoken – the family patriarch pulls up chairs around a barrel and pours out a sample of his beloved vintage. For more certainty, phone ahead.

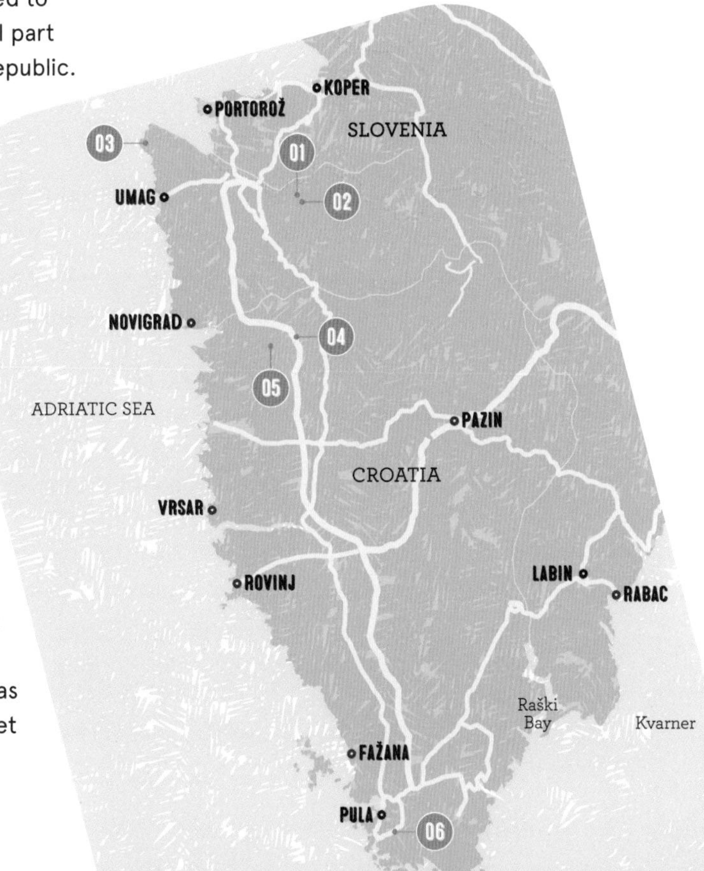

01 VINARIJA KOZLOVIĆ

Set in the greenest of valleys – down the hill and across a gurgling stream from the pretty hilltop town of Momjan – the Kozlović family vineyard has the slickest tasting room in all of Istria. Constructed of stone, glass and angular branch-like metal cladding, it's a striking piece of contemporary architecture, with a slat-roofed dining terrace jutting out over the vine-clad slope. Despite the fancy infrastructure, tastings and tours still need to be arranged in advance – although you can purchase bottles directly from the cellar door anytime during opening hours. The highlight here is the Malvazija Istarska but it also produces excellent Muškat Momjanski; Teran; and a Merlot, Cabernet Sauvignon and Teran blend marketed as Santa Lucia Noir. For something a bit different, try Sorbus, its dessert Muscat, or Mediteran, a port-like Teran fortified with honey.

www.kozlovic.hr; tel 052 779 177; Vale 78, Momjan; Mon-Sat

02 SINKOVIĆ

You may well encounter the unexpected sight of a pig reclining in a wine barrel at this friendly *agroturizam* (farm-based tourism) venture. Say hi to Pepa, the family's utterly charming and totally pampered truffle pig. Located in the teensy village of San Mauro, just up the hill from Momjan, the complex includes rooms, a restaurant, a little shop selling homemade goods from the farm (jams, jellies, balsamic vinegar, olive oil, beer, liqueurs and spirits) and the Sinković winery. Unsurprisingly, it's the Muškat Momjanski that steals the show, but Sinković also makes Malvazija Istarska (in oaked and unoaked versions), Teran, rosé and dessert wine. During the high season, the restaurant and shop are open in the afternoons and evenings; order a flight of wines to enjoy on the terrace, accompanied by local cheese. It's a small operation though, so it's safest to call beforehand, no matter what time of the year you're visiting.

www.sinkovic.hr; tel 052 779 033; San Mauro 157, Momjan; by appointment

01 Istrian coast

02 Seafood and wine by the sea

03 Mountain biking on the Istrian coast

04 Istria's green interior

The welcoming Benazić family turn on an engaging experience at their stylish tasting room on the fringes of Pula, Istria's biggest city

03 DEGRASSI VINA

Tucked away near the sea in Istria's northwesternmost corner, Degrassi produces a surprisingly large range of wines from grapes sourced from vineyards scattered around the surrounding area. At the top of the pile are an excellent Cabernet Sauvignon and a couple of elegant blends: the Terre Bianche Cuvée Rouge (a mix of six red varietals) and the Terre Bianche Cuvée Blanc Riserva (a Malvazija-dominant white blend). Tastings comprise five wines, cheese and charcuterie, served in an atmospheric brick-lined hall with a roaring fire in winter, and there's even a small cellar museum featuring ancient amphorae uncovered from merchant ships wrecked just off the coast. In summertime you can just bowl up to the charming on-site *enoteca* (wine bar and shop), where you can order wines by the glass and snacks to accompany them.
www.degrassi.hr; tel 052 759 250; Podrumarska 3, Bašanija; Mon–Sat Feb–Dec

04 VINA GERŽINIĆ

The friendly folk at this small-scale family vineyard (just 10 hectares/25 acres) are more than happy to welcome visitors, but it pays to call ahead to make sure they're around. They craft an excellent Malvazija Istarska, along with Chardonnay, Teran, Cabernet Sauvignon and Syrah (a relative newcomer to Istria). It's also the place to try Istria's best Muškat Žuti (yellow Muscat), a wine more commonly associated with Italy where it's known as Moscato Giallo and usually produced as a dessert wine. Here it's presented off-dry, perfect for serving chilled as an aperitif or after a meal, with cheese. However wine isn't the only string to the Geržinić bow: its olive oil has been rated

among the top 100 in the world by the prestigious *Flos Olei* guide.
www.gerzinic.com; tel 052 446 285; Ohnići 9; by appointment

05 VINA COSSETTO

There are only 7 hectares (17 acres) under vine at this family vineyard, 12km (7 miles) northeast of Poreč, which tips towards the quality end of the market. One of the grapes grown here is Borgonja, a light red variety thought to have been brought to Istria from France in the 3rd century by Roman emperor Marcus Aurelius. Yes, you guessed right – the word Borgonja is derived from Bourgogne, and the grape in question is actually the familiar Burgundy variety Gamay. Here it's used in a Merlot-dominant red blend, labelled Mosaik. Cossetto also produces Cabernet Sauvignon, Teran, Malvazija Istarska, Chardonnay and white Muscat. Despite the winery's humble size, there's an attractive vaulted-brick tasting room with set opening hours in summer; phone ahead at other times.
www.cossetto.net; tel 052 455 204; Roškići 10, Kaštelir; daily Jun–Aug

06 VINA BENAZIĆ

The welcoming Benazić family turn on an engaging experience at their stylish tasting room on the fringes of Pula, Istria's biggest city. Grab a seat among the trees and potted plants on the polished-concrete terrace and indulge in a flight of six wines (each named after one of the family members), accompanied by bread, cheese and olives. Allow a couple of hours and catch a taxi – these aren't dainty pours. Do try the Katerina Malvazija Istarska, which earned a grand gold medal at Istria's Vinistra wine awards in 2019. Other medal-winners include the Ana Antonija Teran, Veronica Cabernet Sauvignon and Veronica rosé. Although the tasting room has set hours in the summer months, it's still best to call ahead. The shop opens all year, however, and tastings can be arranged with a bit of notice, even in winter.
www.vinabenazic.com; tel 097 798 7643; Valdebečki put 36, Pula; Mon–Sat Jun–Sep

05 Motovun village in Istrian wine country

06 Poreč

ESSENTIAL INFORMATION

WHERE TO STAY

MENEGHETTI WINE HOTEL
Top-quality wine and olive oil are produced at this rural estate near Bale, and the accommodation is just as luxe: stone units set within the vines, adjacent to a historic manor house. There's even a private beach. *www.meneghetti.hr; tel 052 528 800; Stancija Meneghetti 1, Bale; Apr–Dec*

AGROTURIZAM SAN MAURO
Immerse yourself in the winemaking life at this village farmhouse, which rents eight simple rooms with kitchenettes; some have terraces and sea views. Homemade goodies fill the breakfast table. *www.sinkovic.hr/agroturizam; tel 052 779 033; San Mauro 157, Momjan*

WHERE TO EAT

MENEGHETTI RESTAURANT
Sample estate wines while you tuck into exquisite modern Istrian cuisine at one of the region's finest restaurants. It's attached to the swanky Meneghetti Wine Hotel; non-guests are welcome but you'll need to book ahead. *www.meneghetti.hr; tel 052 528 800; Stancija Meneghetti 1, Bale; Apr–Dec*

06

KONOBA DANIELA
Seek out this rustic village tavern, 5km (3 miles) east of Poreč, for traditional Istrian cuisine. The menu includes delicious mushroom ravioli, huge steaks and an excellent steak tartare, as well as cinnamon dumplings with a side of jam. *www.konobadaniela.com; tel 052 460 519; Veleniki 15a, Poreč; daily*

WHAT TO DO

POREČ
With its dazzling 6th-century Euphrasian Basilica, Poreč is a Unesco World Heritage Site filled with glimmering golden mosaics.

ROVINJ
Set on an egg-shaped peninsula, Rovinj is the prettiest town in Istria, its narrow lanes winding up to an immense baroque church.

PULA
This somewhat gritty port city is peppered with relics from its glory days under the Romans. There's a well-preserved 20,000-seater amphitheatre, an elegant triumphal arch and a beautifully proportioned little temple built for the worship of the first emperor, Augustus. A Venetian-era fort at the centre of the Roman street plan is now a history and maritime museum, while an Austro-Hungarian fort by the beach has been cleverly converted into an aquarium. Also interesting are the gorgeous Secession-style central market building and the 13th-century Franciscan Monastery.

CELEBRATIONS

In May, Poreč hosts the Vinistra wine fair, during which Istria's wine industry dishes out medals for the best vintages. Then, on the last Sunday of the month, comes Open Wine Cellar Day, when vineyards across the region throw open their gates. But the biggest feast for winemakers across Croatia and Slovenia is Martinje (St Martin's Day, 11 November), with the sampling of new vintages.

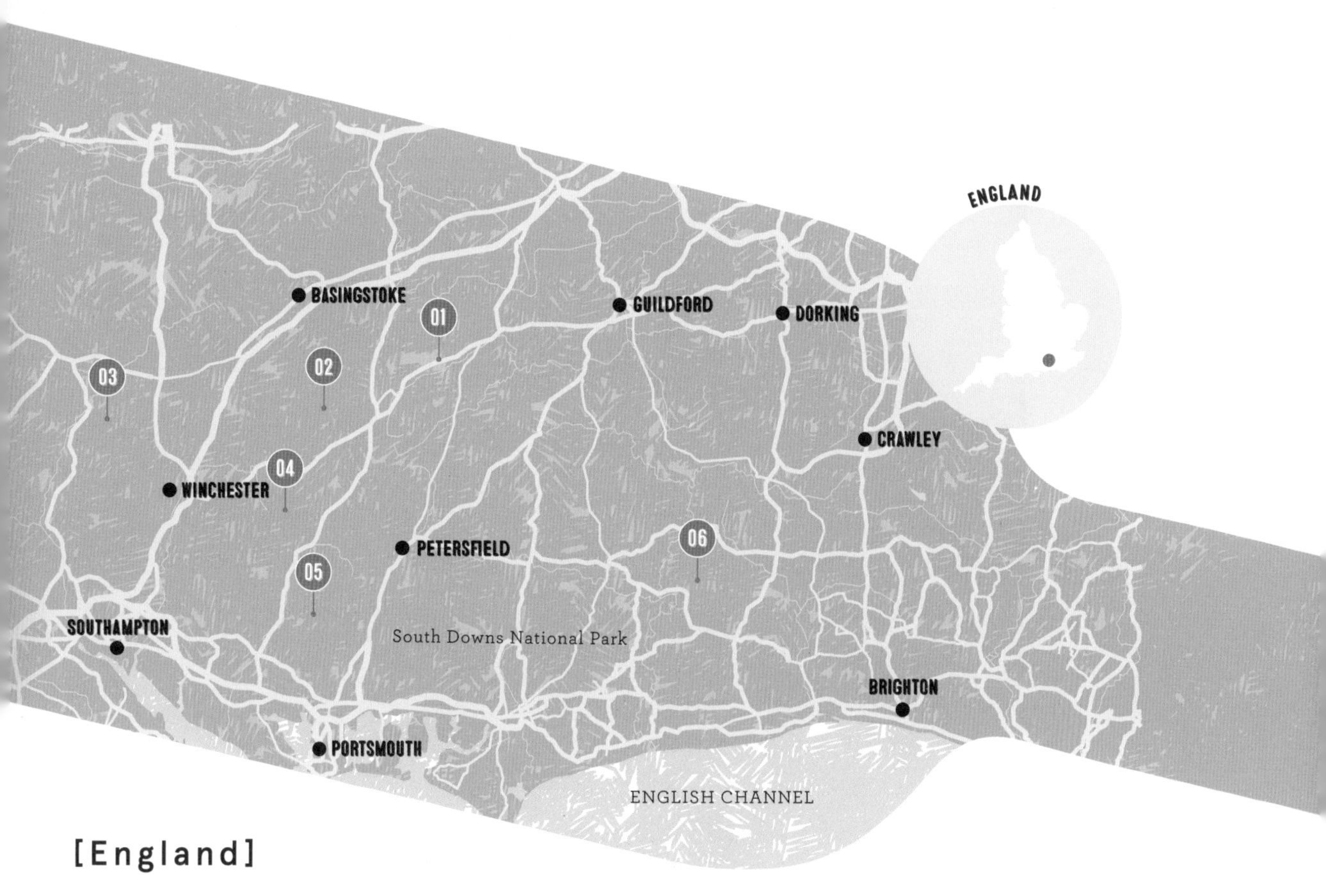

[England]

SOUTH DOWNS

It's fizzy, refined and winning awards: English white wine sparkles in the summer, the perfect time to take a tour of Southern England's vineyards.

Overhead a skylark sings in the blue sky. Green fields sweep down from a chalk ridge laced with white tracks. To the south lies the sea, to the north the counties of Hampshire and Sussex. These are England's South Downs in summer, a place of villages, hiking trails and, increasingly, vineyards. For the South Downs, now a National Park, are a narrow, 160km-long (100 miles) spine of chalk hills that run southeast all the way from Winchester, an ancient capital of England, to Eastbourne; the same seam of rock re-emerges across the Channel in Champagne country.

English wine was long a laughing stock, not least among the French, being too thin, too sour or over-sweetened. But in the last 15 years, the South Downs region has been the source of some excellent Champagne-style sparkling wines. In truth, England's wineries are spread out over quite a distance, from Tenterden in Kent to Three Choirs in Gloucestershire, and a tour taking in all of them would be impracticable. They don't all lie on the South Downs; many are further inland, and few are open to the public (so far – that will change). But there are a handful of vineyards concentrated in tranquil Hampshire that together make a weekend exploring English wine, and some of the county's other attractions, enjoyable and, perhaps, something of a revelation.

GET THERE
Southampton and London Gatwick are the closest airports but the region is only an hour from London by train.

01 JENKYN PLACE

'We were at an event at Nyetimber in 2004,' recalls Simon Bladon, 'and the wine was very good. I asked them where it was from and they said "here". So I came back and planted a field of vines, as you do.' Jenkyn Place is on the northern edge of the chalk band, with the Downs about half an hour's drive south. This is a very pretty corner of Hampshire, close to Chawton where author Jane Austen lived. Jenkyn Place is only open to the public from June to September each year, as part of the Hampshire Fare food festival – but summer is the best time to visit anyway.

www.jenkynplace.com; Hole Lane, Bentley; open days Jun–Sep

02 HATTINGLEY VALLEY

With winemaker Emma Rice – a two-time English winemaker of the year – at the helm, Hattingley Valley is one of the most highly regarded wineries in the country. The vines were established in 2008 and the state-of-the-art winery itself was completed in 2010, harnessing solar power and science in its own laboratory. Add Emma's experience in Tasmania and California and Hattingley's sparkling wines get a stellar start in life. They live up to their promise too, as a tasting after the winery tour in this relatively remote part of north Hampshire proves.

www.hattingleyvalley.com; tel 01256 389 188; Wield Yard, Lower Wield, nr Alresford; by appointment year-round except during harvest $

03 COTTONWORTH WINES

Surrounded by the clear, braided streams of the Test Valley, which flow over chalk beds and form watercress-filled meadows, and the thatched cottages of such villages as Wherwell, Cottonworth Wines' setting couldn't be more English. During the past 10 years, the Liddell family has planted the sparkling wine grapes of Chardonnay, Pinot Noir and Pinot Meunier on south-facing slopes around their farm. But the weather can be as English as the setting. 'Our biggest challenge in the Test Valley is a short growing season in a cool climate,' says

01 South Downs National Park

02 Vineyards at Raimes

03 Simon Bladon and his daughter Camilla Jennings of Jenkyn Place

04 South Downs village life, Lewes

Federico Firino of Cottonworth. 'We need to be lucky enough in order to get the right weather at the right time. We are at the mercy of Mother Nature!' When everything comes together, though, there's no better way of enjoying Cottonworth's wines than with some of the local smoked trout overlooking the vineyard on a sunny afternoon. Federico believes that visiting English wineries is a great way of introducing the wine, 'but we didn't want to make it too exclusive like many producers might do in Champagne.'

England being a new wine region means that winemakers like Hugh Liddell don't feel constrained by history or tradition. However, quality English wine already has PDO (Protected Designation of Origin) status and some believe that regions such as the South Downs may apply for a system like France's *Appellation d'Origine Contrôlée* (AOC).

www.cottonworth.co.uk; tel 01264 860 531; Cottonworth House, Andover; tours Fri–Sat Jun–Aug

04 RAIMES

When you pass through the hamlet of Tichborne (essentially just a pub beside the Itchen River) look for a sign on your right, next to a farm track. This is Raimes, where owners Augusta and Robert Raimes have farmed for five generations. In 2011 they planted Chardonnay, Pinot Noir and Pinot Meunier vines on south-facing slopes and you can now taste the fruits of their labour on summer weekends at the farm's cellar door. Some of the sparkling wines have won awards, such as a gold medal at the International Wine Challenge in 2019 for the Classic. But for Augusta Raimes: 'The most rewarding part has been adapting to a hand-grown and hand-picked crop – and it's also a great pleasure to introduce people to English wine and see them realise how good it is.'

Although the grapes are grown along the same chalk spine that stretches to Champagne in France, some tweaking of the technique has had to be made at Raimes. 'We establish a tall trunk to wire height which helps keep the buds

away from the cold ground at tricky springtime when there can be ground frosts,' says Augusta. *www.raimes.co.uk; tel 01962 732 120; Grange Farm, Tichborne; tours by appointment Jul–Sep, tastings (with fee) by appointment year-round, free tastings Fri, Sat, Sun May–Sep*

05 HAMBLEDON VINEYARD

Hambledon Vineyard is set in the idyllic Hampshire village of the same name: a place of hills, fields, woods and little-used flint-strewn lanes. It's been 10 years since Chardonnay vines were planted here. 'The chalk on which we grow our vines was formed on the seabed of the Paris basin some 65 million years ago,' says managing director Ian Kellett. 'The same chalk is found in the best Chardonnay areas of the Côtes des Blancs in Champagne.' With a duo of winemakers on board – Hérve Jestin of Champagne Duval Leroy, and Felix Gabillet, a graduate of Changins School of Wine in Switzerland – Hambledon's European connection is strong. Indeed, in 2015, Hambledon's Classic Cuvée beat French Champagnes in a blind tasting hosted by *Noble Rot* wine magazine, thanks to its scents of 'fresh sourdough, magnolia and lily with a hint of smoke'. A new tasting room is scheduled for 2020. *www.hambledonvineyard.co.uk; tel 02392 632 358; Hambledon; by appointment* $

06 NUTBOURNE VINEYARD

Cross the border into West Sussex to visit Nutbourne Vineyard. Nearby Nyetimber may have been in the vanguard of the English sparkling wine revolution – but it's not open to the public. However, Nutbourne in Pulborough is open to all, and its white wine was the first English still wine to win a gold medal at the International Wine and Spirit Competition. The still wines use Riesling-style grapes to create fruity, aromatic flavours, typically with a hint of elderflower. The sparkling wines use Pinot Noir and Chardonnay. Discover the process with a tutored wine tasting in the vineyard during summer months.

The family-owned winery is based in a 19th-century windmill midway along the Downs. From here you can return to London or cities such as Brighton or Chichester – or just linger in the countryside for walks. *www.nutbournevineyards.com; tel 01798 815 196; Gay Street, Pulborough; by appointment, usually Sat May–Oct* $

05 Jenkyn Place vineyards

06 Surveying the beauty of the South Downs

ESSENTIAL INFORMATION

WHERE TO STAY

HOTEL DU VIN

For city-centre accommodation and a well-crafted wine list, try the Hotel du Vin in Winchester, gateway city at the west end of the South Downs. Part of a chain that prides itself on its wine, there's another Hotel du Vin in Brighton at the opposite end of the South Downs.
www.hotelduvin.com

THE ANGEL INN

In the centre of the South Downs town of Petworth, the Angel Inn offers cosy (if not cheap) accommodation close to the town's antique shops and 17th-century Petworth House, which has a remarkable art collection, wood carvings by Grinling Gibbons and a deer park designed by Capability Brown.
www.angelinnpetworth.co.uk; tel 01798 344 445; Angel Street, Petworth

WHERE TO EAT

THE HAWKLEY INN

Deep in Hampshire's steep hills (known as 'hangers'), south of Jenkyn Place and Hattingley vineyards, this country pub serves superb lunches and local cask ales, which are best enjoyed in the large garden if the weather permits. Walk off your lunch in the woods.
www.hawkleyinn.co.uk; tel 01730 827 205; Pococks Lane, Hawkley

THE FORTE KITCHEN

Hampshire's produce stars on the breakfast and lunch menus at this stylish space in central Winchester.
www.thekitchenrestaurants.co.uk/fortekitchen; tel 01962 856 840; Parchment Street, Winchester

WHAT TO DO

SOUTH DOWNS NATIONAL PARK

The South Downs have long attracted hikers, mountain bikers and horse riders. The South Downs Way trail runs, up and down, for 160km (100 miles) from Winchester to Eastbourne but is accessible at lots of points, so you can easily stretch your legs a section at a time.
www.southdowns.gov.uk

LANGHAM BREWERY

Take a break from the grape with a visit to this brewery in the South Downs National Park, just west of Pulborough. Superb golden ales, such as the summery Sundowner, are brewed here. Tours are possible and there's a shop for quick purchases.
www.langhambrewery.co.uk; tel 01798 860 861; The Granary, Langham Lane, Lodsworth

CELEBRATIONS

FIZZ FEST

Fizz Fest sees the seven members of the Vineyards of Hampshire come together for an annual celebration of sparkling wine, usually in July. Not all of the vineyards are open to the public so it's a great opportunity for some comparative tastings plus masterclasses, music and local food.
www.vineyardsofhampshire.co.uk/events

HAMPSHIRE FOOD FESTIVAL

Indulge in local produce from across the county – buffalo mozzarella from near Stockbridge, gin from Winchester – during this July celebration. Wine events include pop-up suppers and tasting tours.
www.hampshirefare.co.uk

01

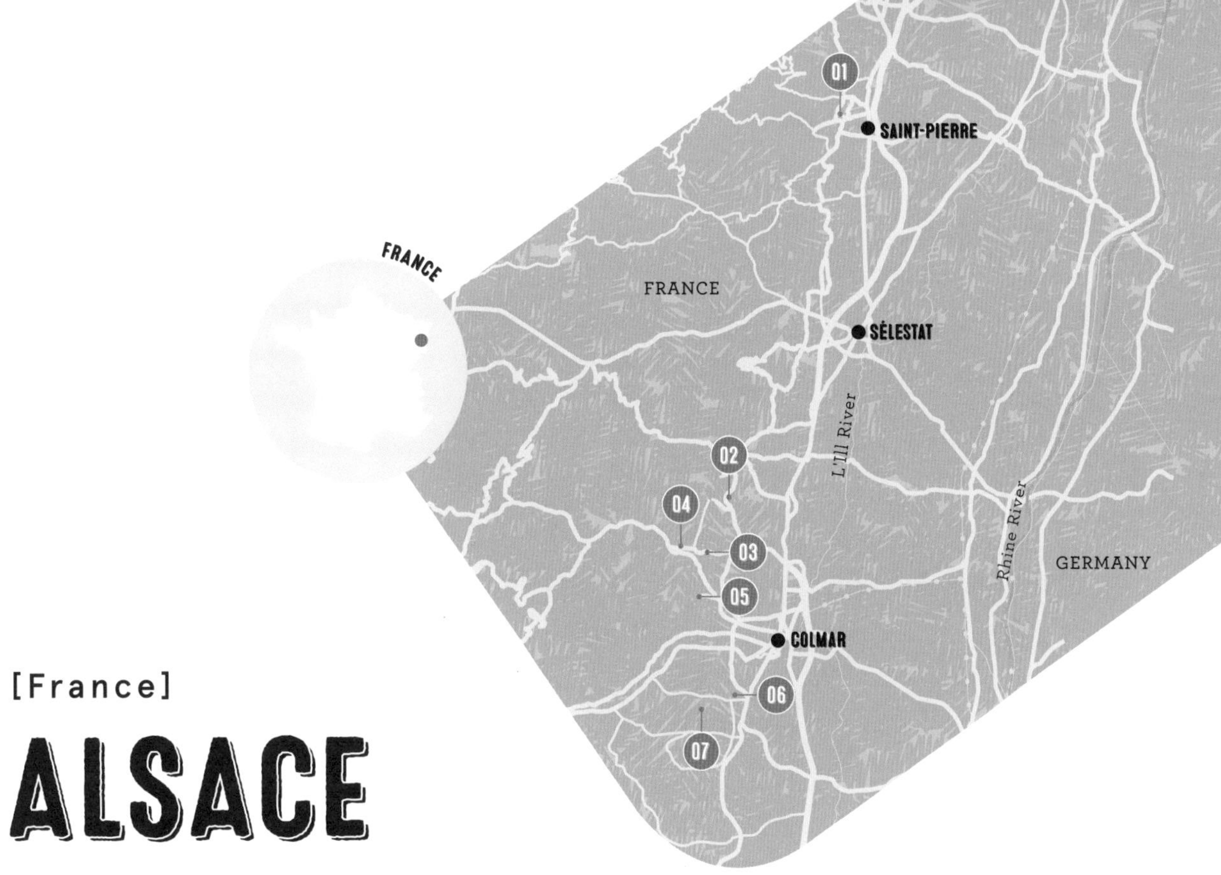

[France]

ALSACE

Take a trip through the picturesque villages of northeast France to sample distinctive white wines among traditional, timbered wineries.

Winemaking in Alsace has had its ups and downs throughout a long history, which begins with the Romans planting the first grapes. In the Middle Ages, records show 160 villages growing vines, while by the 16th century, they were making some of the most prized wines in Europe. Then came 300 years of war, phylloxera and political ping-pong, with the territory passed back and forth between France and Germany. A renaissance has taken place over the last 50 years, seeing a significant increase in quality, and today Alsace is in the vanguard of the movement towards organic cultivation. The picture-perfect vine-clad hillsides here are the ultimate terroir for white wines, with Pinot Noir the only red among the seven official grape varieties. Each village, each winemaker, creates a complex patchwork of interpretations of Riesling, Gewurztraminer, Pinot Blanc and Gris, Muscat and Sylvaner.

Nowhere in France can compare when it comes to the welcome given to wine tourists. Alsace was one of the first regions to organise its own Route du Vin, some 60 years ago, and today local *vignerons* are always coming up with new ideas to attract winelovers: bike tours and marathon races through the vines; food and wine fairs; a procession honouring Alsace's iconic Gewurztraminer past the fairy-tale half-timbered houses of Bergheim (a strong candidate for the most beautiful village in France); and night-time illuminations of medieval churches as bottles of bubbly Crémant are popped in celebration. And throughout Alsace there is a long tradition of winemakers opening up their rustic houses as welcoming *chambres d'hôtes*. Rather than a formal tasting, guests are often privileged to sit down with the winemaker for a relaxed session trying some favourite vintages; a long aperitif that often continues over dinner in a local bistro.

GET THERE

EuroAirport Basel Mulhouse Freiburg is the nearest major airport, 99km (62 miles) from Mittelbergheim. Car hire is available.

01 DOMAINE ALBERT SELTZ

Albert Seltz is a 16th-generation winemaker and Alsace's champion of the humble Sylvaner, which he describes as 'the grape no one wants to talk about and prefers to imagine does not exist'. Sylvaner is often overlooked compared to the likes of Riesling, and Seltz went on a crusade to make French officialdom recognise the Sylvaner vines around here as a Grand Cru. Visitors can sit back as Albert takes them through a dozen of his sensational Sylvaner vintages, from the wonderfully drinkable Sylvaner de Mittelbergheim through to a seductive 2013 'Sono Contento' Vieilles Vignes, which he poetically describes: 'Look at the colour, a dull gold that is autumn. Smell the nose; now imagine that with sautéed wild trout.'

www.albert-seltz.fr; tel 03 88 08 91 77; 21 Rue Principale, Mittelbergheim; daily by appointment $

02 DOMAINE BECKER

An old barn of the Beckers' rambling winery has been converted into a giant *winstub* (traditional Alsatian 'wine lounge'), with an ambience more like a jolly pub than a sophisticated wine bar. In addition to her own organic wines Martine Becker also promotes local specialities: organic honey and jam, foie gras and distinctive Alsace pottery. Martine is a mine of legendary village tales: 'Can you imagine that during WWII over 150 of the villagers used to sleep down in our cellar at night to avoid bombs – and Papa said they also drank a lot of our stock to avoid the occupying German army getting it!'

www.vinsbecker.com; tel 03 89 47 90 16; 4 Route d'Ostheim, Zellenberg; daily Apr–Dec, by appointment at other times

03 DOMAINE PAUL BLANCK ET FILS

Sitting in the rustic wood-panelled tasting room of Domaine Blanck's 16th-century cellar, surrounded by vast oak casks painted with traditional Alsatian scenes, it's quickly apparent that Philippe

01 Colmar's Little Venice district

02 Harvest at Maison Emile Beyer

03 Fine reds from Domaine Weinbach

04 Gewurztraminer grapes, Domaine Weinbach

Blanck is an expert at explaining the complex world of Alsace wines, from a simple Sylvaner to a Grand Cru Pinot Gris. Philippe offers advice that is appropriate for all Alsace cellar guests: 'Visitors here should do three things: see our fabulous vineyard landscapes, try the wine with the winemaker himself, then ask him where to eat, as our wines are best when you taste Alsatian cuisine at the same time.'
www.blanck.com; tel 03 89 78 23 56; Grand'Rue, Kientzheim; Tue–Sat $

04 DOMAINE WEINBACH

The Route du Vin that runs into Kaysersberg, another idyllic Alsatian village, is marked by a long stone wall protecting an ancient vineyard and Capuchin monastery, now a grand mansion and winery, where Catherine Faller and her son Théo make a sensational selection of wines. Behind the wall is the ancient Capuchin monastery where monks planted the surrounding 5-hectare (12-acre) Clos des Capucins in 1612. The monks were evicted during the French revolution and Catherine's grandfather bought the property in 1898.

While tasting vintages such as the complex, concentrated Pinot Gris Altenbourg 2018 or a luscious late-harvest Gewurztraminer, Catherine is full of suggestions for food pairings. 'Can't you imagine this Muscat with fresh asparagus, the Pinot Blanc with a cheese soufflé,' she enthuses, 'or our full-bodied Pinot Noir S de Schlossberg alongside a succulent leg of lamb?'
www.domaineweinbach.com; tel 03 89 47 13 21; 25 Route du Vin, Kaysersberg; Mon–Sat by appointment $

05 VIGNOBLE KLUR

The Klur family have created a paradise getaway for eco wine-lovers. While Clément Klur produces an outstanding selection of biodynamic wines, he has also transformed his family mansion into a bohemian B&B, offering wine courses, traditional Alsatian

05 Riquewihr, near Colmar

06 Traditional Alsace *Kugelhopf*

07 Ballons des Voges forest

cooking classes and even poetry readings. The grounds extend over vegetable gardens, ponds, a sauna and a solarium, and walking tours are organised through the stunning terraced vineyard looking out over Katzenthal – Valley of the Cats. Klur is innovative in the cellar, too, refusing to add any sulphites, so the whole range here is now classed as 'natur' (natural), including a striking orange wine blending Gewurztraminer with Muscat grapes.
www.klur.net; tel 03 89 80 94 29; 105 Rue des Trois Epis, Katzenthal; Mon–Sat $

06 MAISON EMILE BEYER

Visitors flock to enchanting Eguisheim, unchanged since the Middle Ages, gathering in the town square to stare up at the iconic storks' nests balancing atop the church steeple. One of the traditional, brightly coloured half-timbered houses on the square has been the winemaking home of the Beyer family since 1580. Today they have a modern winery on the outskirts of town, but tastings are held here in the cobbled courtyard. As is the case all over Alsace, be prepared for a marathon session, as the Beyers produce some 30 different wines on their 16-hectare (40-acre) organic estate. Try a surprisingly dry Muscat or the full-bodied oak-aged Pinot Noir, though the stars of the show are the Grand Cru Rieslings. While Maman Beyer holds court serving the wine, her son, Christian, a 14th-generation winegrower, notes, 'Records exist that grapes were first grown on these rolling hillsides by the Romans, and I believe when you are standing among the vines that produce our Riesling Grand Cru Eichberg – Hill of Oaks – you absorb this incredible history and heritage.'
www.emile-beyer.fr; tel 03 89 41 40 45; 7 Place du Château, Eguisheim; Mon–Sat

07 GERARD SCHUELLER

This 500-year-old winery has no tasting room, just a rickety table wobbling with a dozen half-opened bottles. It's squeezed between steel vats and barrels, and a workspace where labels are stuck – by hand – onto magnums of the highly original Bulle de Bild, a sparkling Gewurztraminer blended with 10% Riesling grape juice, which provocative Monsieur Schueller refuses to call a Crémant. It turns out that he often falls foul of the authorities with his unorthodox 'natural wines' failing to pass official tasting tests. 'I like to leave my wines open when tasting just to see whether there is an effect of oxidisation. But I'm not scared, and they only seem to get better the longer they are open,' he explains. And this is borne out when the tasting begins, as he digs out a bottle opened two weeks ago of a wonderful lush, dark-amber 2008 Riesling Grand Cru.
Tel 03 89 49 31 54; 1 Rue des 3 Châteaux, Husseren les Châteaux; Mon–Fri by appointment

ESSENTIAL INFORMATION

WHERE TO STAY

CLOS FROEHN
Guests are pampered at Martine and Alphonse Aubrey's 17th-century cottage, which overlooks the vineyards. At breakfast, Alphonse (formerly the village baker) serves cakes and pastries.
www.clos-froehn.com; tel 03 89 47 95 68; 46 Rue du Schlossberg, Zellenberg

SYLVIE FAHRER
The rooms are simple in this reasonably priced B&B, but breakfast is in a grand half-timbered salon. Evening wine tastings are held in a converted barn filled with barrels and tractors.
www.fahrer-sylvie.com; tel 03 89 73 00 40; 24 Route du Vin, Saint-Hippolyte

WHERE TO EAT

FERME-AUBERGE DU KAHLENWASEN
High above the wine village of Turkheim, enjoy panoramic views and hearty home cooking that makes use of wonderful fresh farm produce: smoked meat; potatoes fried with bacon; Munster cheese made just that morning.
www.facebook.com/Kahlenwasen; tel 03 89 77 32 49; Luttenbach-près-Munster

CAVEAU MORAKOPF
You won't see many tourists in this snug, welcoming bistro, but locals tuck into enormous portions of *jambonneau* (crispy pork knuckle) or delicious goose foie gras.
www.caveaumorakopf.fr; tel 03 89 27 05 10; 7 Rue des Trois Épis, Niedermorschwihr

BRASSERIE L'AUBERGE
Colmar is the wine capital of Alsace, and this atmospheric century-old brasserie is a temple to traditional cuisine, serving up steaming plates of tangy *choucroute* (sauerkraut dressed with sausages).
www.grand-hotel-bristol.com/en/restaurants/brasserie-lauberge; tel 03 89 23 17 57; 7 Place de la Gare, Colmar

WHAT TO DO

BALLONS DES VOSGES
This national park is a paradise for nature-lovers, with a rolling range of pine-clad peaks, lush valleys and tranquil lakes. During summer it's popular for hang-gliding, rambling, canoeing and, above all, cycling – the park is often featured during the Tour de France. In winter, visitors come for the superlative cross-country skiing, as well as small, family-friendly downhill resorts.
www.parc-ballons-vosges.fr

CELEBRATIONS

Molsheim is home to a memorable weekend festival in June, which includes a half-marathon run through the vineyards.
www.marathon-alsace.com

01

FRANCE

[France]

BORDEAUX

Discover another, more accessible side to world-famous Bordeaux: in the saddle, in the trees or up in the air.

Premier Cru classé; Left Bank, Right Bank; €6000 cases of Château Margaux: the world of Bordeaux wine can seem an intimidating, confusing and, yes, expensive place. For one thing, France's largest *Appellation d'Origine Contrôlée* (AOC) comprises several subregions, each very different. On the Right Bank of the Dordogne River lie Saint-Émilion and Pomerol. On the Left Bank of the Garonne River, which flows through the city of Bordeaux and meets the Dordogne in the vast Gironde estuary, is the Médoc, home to Margaux and other hallowed names.

But one region of Bordeaux has made a special effort to be more accessible than its neighbours: Graves and Sauternes, upriver of Bordeaux on the Left (south) Bank of the Garonne. During the last ten years, wineries along the Graves and Sauternes wine route have introduced ever more novel ways of tasting and learning about the region and its unique wines: visitors can cycle to châteaux, go canoeing, take sightseeing flights from vineyards and taste wines in tree houses.

This trail begins and ends in the dynamic university city of Bordeaux, whose fortunes have ebbed and flowed like the broad Garonne that bisects it. Grapes were first grown here during the region's Roman period when Burdigala was an import/export hub and there are still vineyards within the city limits. Come the 12th century, Eleanor of Aquitaine's marriage to Henry II in 1152 enabled Aquitaine to become England's sole supplier of wine. In the 16th century, the Dutch encouraged the Bordelais to expand. By the 18th century, wine merchant families such as the Lurtons and the Bartons ushered in Bordeaux's next golden age, building many of its most beautiful châteaux.

GET THERE
Bordeaux has an international airport on its outskirts (take bus 1 to the centre for €1) and is 2hr from Paris by train.

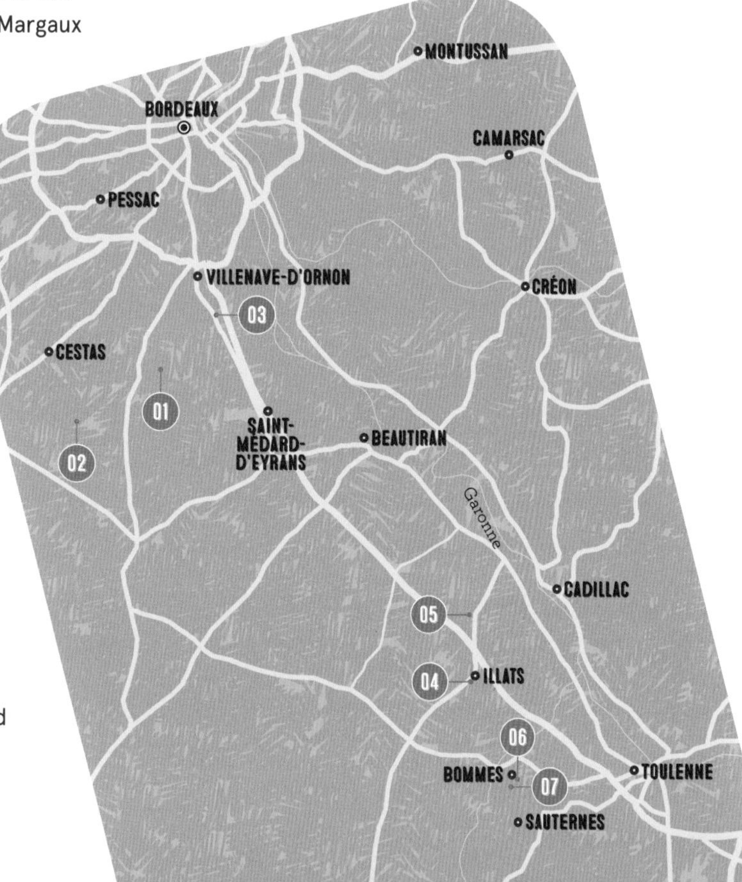

01 CHÂTEAU LARRIVET HAUT-BRION

Rouse your taste buds with a chocolate and wine tasting at Château Larrivet Haut-Brion, a few minutes' drive south of Bordeaux in the Graves region's Pessac-Léognan appellation. Elsewhere on this trail you'll try dry red wines in Graves and sweet white wines in Sauternes; red wines at Larrivet Haut-Brion are paired with single-origin chocolates from Bordeaux's Chocolaterie Saunion.

Since the 19th-century château isn't one of Bordeaux's Premier Crus (the five prestigious 'First Growth' châteaux – Lafite, Latour and Mouton in Pauillac, Margaux, Haut-Brion in Graves – received their designation in 1855) it's free to be more experimental, according to Larrivet Haut-Brion's wine guide Alexandra Monfort. For example, white wines are fermented in giant concrete eggs. Red wines, however, remain in French oak barrels for six to 18 months. 'The collaboration between a cooper and a winemaker is crucial,' says Alexandra. 'A light toast imparts flavours of vanilla and cinnamon and a darker toast brings stronger flavours such as tobacco.'

The owners, the Gervoson family, host chocolate-egg hunts in the château's Jardin d'Ivresse around Easter time.

www.larrivethautbrion.fr; tel 05 56 64 75 51; 84 Avenue de Cadaujac, Léognan; Mon–Fri by appointment $

02 CHÂTEAU DE LÉOGNAN

The emblem on the bottles of Château de Léognan's Cabernet Sauvignon and Merlot blend depicts a phoenix, a symbol borrowed from the chapel in the grounds of this country house. It's an appropriate choice: since Philippe and Chantal Miecaze purchased this beautiful rural property on the west side of Léognan in 2006 they've restored its every corner. The main house now features several B&B suites, and a restaurant, Le Manège, occupies the tree-shaded horse-training ring.

Château de Léognan is part of the Pessac-Léognan appellation, which was founded by André Lurton in 1987 and is the only Bordeaux appellation outside Médoc to have a Premier Cru and to be part of a city. According to Chantal Miecaze, Bordeaux wine has a natural advantage: 'The soil in Graves is well-drained because it used to be an estuary,' she explains. 'Before settling on its final course, the Garonne river

01 The Garonne at Bordeaux

02 Take to the skies above Château Venus

03 Wine tours on wheels at Château Bardins

meandered from the Pyrenees to the Atlantic, bringing gravel down from the mountains. The gravel is great for grapes: water drains quickly so the vines have to make more of an effort to seek it, and the pebbles reflect the sun, lengthening the day for grapes to ripen.' You can sample the wine over lunch or at a tasting.
www.chateauleognan.fr; tel 05 56 64 14 96; 88 Chemin du Barp, Léognan; Mon–Sat by appointment

03 CHÂTEAU BARDINS

At the next stop, Château Bardins, Pascale Larroche and fifth-generation owner Stella Puel offer a very active approach to discovering Graves' wines. Formerly a farm and still enjoying a pastoral setting, Château Bardins has been producing wine since WWII, in addition to honey, preserves, crops and livestock. Visitors can take an hour's bicycle ride around the locale with Pascale.

'We take in 10 châteaux in 10km [6 miles],' she says. 'That's only possible in Bordeaux where vineyards are so close together.' Along the way Pascale explains why the ground is good for vines: 'they only get a little to eat and drink – their roots need to go down 7m [23ft] sometimes to find water – and annual *épamprage*, or pruning, stops them wasting energy growing shoots that won't fruit.' While the vines may go hungry, there's no need for visitors to: Stella is happy for people to bring picnics to enjoy in the château's wooded grounds, especially if you buy some of the organic wine.
www.chateaubardins.fr; tel 05 56 30 78 01; Chemin de la Matole, Cadaujac; by appointment $

04 CHÂTEAU JOUVENTE

Head southeast down the back roads to the centre of the town of Illats and Château Jouvente, where you'll find some of the best-value red wine in Graves. This is a small, new winery owned by Benjamin Gutmann who contracts the winemaking out to Oliver Bernadet. A one-hour tour – with quiz questions for families – shows how the wine is produced before you taste it. The white is a classic Graves – fruity but elegant: 'peach and pineapple from the Semillon and citrus from the Sauvignon,' says Benjamin. The red is a characterful blend of Petit Verdot, Merlot and Cabernet Sauvignon and at €14 it's a bargain.
www.chateau-jouvente.fr; tel 05 56 62 49 69; 93 Le Bourg, Illats; Mon–Fri, by appointment Sat–Sun

05 CHÂTEAU VENUS

Bumping down a grassy runway between rows of vines in a truly tiny two-seater plane is an exciting way to start your tour of Château Venus. Once airborne, pilot (and winemaker) Bertrand Amart will take you on a looping 20-minute circuit of Sauternes and Graves;

04 Bordeaux's Cité du Vin museum

05 Interactive displays inside the Cité du Vin

longer flights up to Saint-Émilion and Pomerol or all the way to Le Bassin d'Arcachon, weekend beach retreat for the Bordelais, are also possible (prices range from €69 to €189, depending on the route). Even on the short flight you can see all the way to the Dune of Pilat on the coast and get a good understanding of the geography of the Garonne. And on the half-day Toutes Options itinerary (€399), you'll combine all seven flights on offer.

Bertrand started flying lessons at the age of 18. His other passion is winemaking – he worked with his wife Emmanuelle in Napa and Barossa before returning to start Château Venus. They produce sustainably farmed red wine that is designed to be both fruity and immediate.
www.chateauvenus.com; tel 06 03 17 91 39; 3 Pertigues Lieu dit Brouquet, Illats; Mon–Sat

06 CHÂTEAU SIGALAS RABAUD

Château Sigalas Rabaud is in the heart of Sauternes, the region where France's most famous dessert wine is produced. A near neighbour across the fields, Château Yquem, is opening up to the public, but it costs from €75 to €200 for a private tasting there. Sigalas-Rabaud is a more accessible option, thanks to owner and director Laure de Lambert Compeyrot, the sixth generation of her family to run the property. She has opened B&B suites overlooking the vines and offers a range of tastings, from two-hour sensory workshops to family-friendly explorations of the vineyard. Laure explains why this pocket of land is perfect for creating golden, sweet Sauternes: 'The Ciron River that flows nearby is fed from a spring and is very cold. The temperature difference in summer and autumn brings fog to the surrounding vineyards, which is good for causing the botrytis fungus on the grapes that we need to make Sauternes.' This 'noble rot' dehydrates the grapes and adds a honeyed flavour.
www.chateau-sigalas-rabaud.com; tel 05 57 31 07 45; Bommes; by appointment $

07 CHÂTEAU DE RAYNE VIGNEAU

At Château de Rayne Vigneau, one of the 1855 Grand Cru classé châteaux, winemaker and managing director Vincent Labergere is set on updating Sauternes to modern tastes. 'The 19th century was a golden age here,' he says. 'Then Rayne Vigneau's fame slipped away.' Investment from new owners has reignited ambition. 'I wanted to make a more thirst-quenching, versatile wine,' explains Vincent, 'so we brought a new freshness to the wine. We're looking for more acidity and less sugar now.' Similarly, he is intent on introducing the wine and the region through new experiences: you can take a horseback tour, taste the wines in a treehouse, blend your own wines or be a grape-picker for a day (followed by dinner with the winemaker).
www.raynevigneau.fr; tel 05 56 76 64 05; 4 Le Vigneau, Bommes; daily Apr–Nov, Mon–Fri Dec–Mar $

ESSENTIAL INFORMATION

WHERE TO STAY

DOMAINE ECÔTELIA

A fun spot for families: choose from yurts, log cabins, safari-style lodges and even spacious tree houses. Each has abundant space and there's also a swimming pool and other areas to explore. Breakfasts are provided each morning and healthy meals are available at other times. *www.domaine-ecotelia.com; 05 56 65 35 38; 5 Lieu dit Tauzin, Le Nizan*

MERCURE BORDEAUX CHATEAU CHARTRONS HOTEL

This hotel in the former wine merchant district of Bordeaux offers decent-value accommodation close to the Cité du Vin, the riverside attractions and the restaurants of Bordeaux's old town. *www.accorhotels.com; 05 56 43 15 00; 81 Cours Saint-Louis, Bordeaux*

WHERE TO EAT

LE PETITE GUINGETTE

This hip open-air tapas bar serves local wines, beers and snacks from a shack in a Sauternes courtyard from spring to autumn. It's a relaxed local favourite, where vineyard-owners and workers socialise. Check the Facebook page for details of upcoming musical performances. *www.lapetiteguinguette.com; 2 Chemin de Pasquette, Sauternes*

05

HÔTEL-RESTAURANT LALIQUE

At the other end of the spectrum, this stately restaurant in Premier Cru classé Château Lafaurie-Peyraguey offers a good-value set three-course lunch menu with matched wines for €65. It won its first Michelin star in 2019. Wine tastings and accommodation also. *www.lafauriepeyragueylalique.com; tel 05 24 22 80 11; Lieu dit Peyraguey, Bommes*

WHAT TO DO

CHÂTEAU DE ROQUETAILLADE IN MAZÈRES

Take a vineyard breather with a history lesson from Sébastien de Baritault at Château de Roquetaillade, which has been passed down through his family for 800 years. It was based on an 8th-century castle then rebuilt in the 14th century following the design of Welsh castles. Although the exterior is fortified, the interior was restored by French architect Viollet-le-Duc in the 19th century and is a fascinating display of Arts and Crafts design inspired by William Morris and William Burges' Gothic Revival work.

CITÉ DU VIN

The Cité du Vin offers a cultural context to the world of wine with temporary exhibitions and a permanent floor of excellent interactive displays plus restaurants, a wine cellar, a library and a top-floor viewing gallery – all housed in a dramatic building. Arrive by boat to get a sense of the wealth generated by Bordeaux's wine trade as you pass the 19th-century waterfront mansions. *www.laciteduvin.com; www.infotbm.com*

CELEBRATIONS

BORDEAUX FÊTE LE VIN

Every two years this four-day festival, dedicated to the wines of Bordeaux and Aquitaine, takes over the city's riverfront with performances, tastings, workshops and food. *www.bordeaux-wine-festival.com*

[France]

BURGUNDY

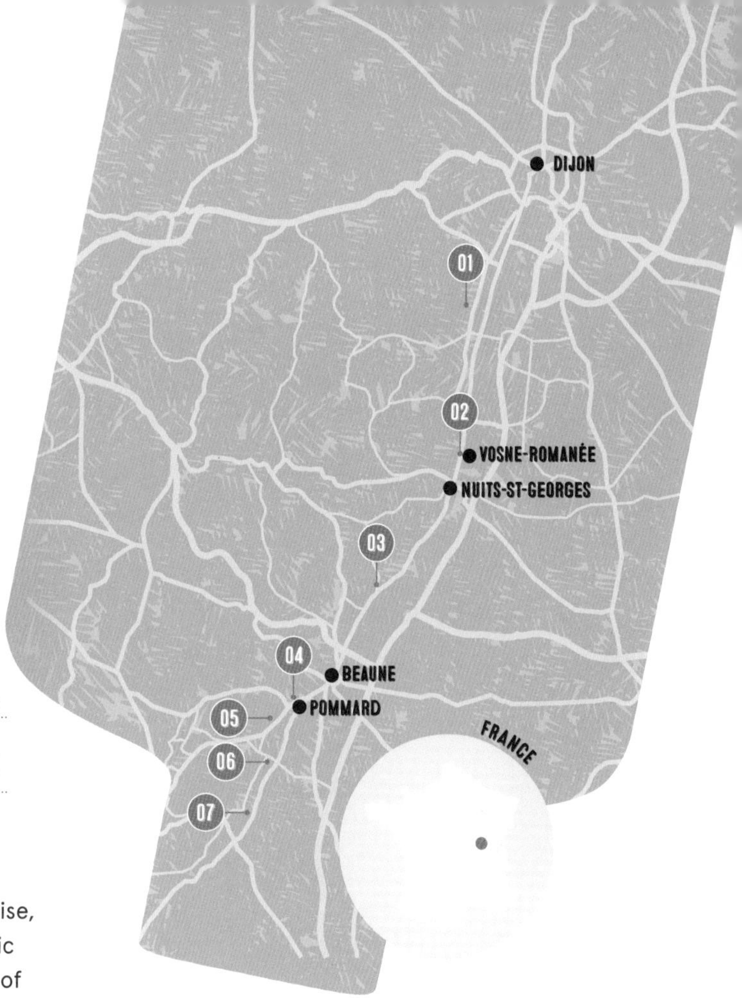

It's easy to feel daunted by Burgundy's reputation, but don't – the locals love to share the secrets of their legendary Pinot Noir with visitors.

Burgundy stretches as far as the vineyards of Chablis, Mâcon and the Côte Chalonnaise, but the quintessential heart of this historic winemaking region is the 80km (50-mile) stretch of road along the Côtes de Nuits and Côtes de Beaune, from Dijon down to Santenay. The illustrious vineyards that line both sides of La Côte d'Or, or Gold Coast, as it is known, produce probably the most famous wines in the world. This essentially monoculture terroir of painstakingly manicured vines provides the perfect interpretation of Pinot Noir and Chardonnay, grapes that may be grown all over the rest of the world but which attain unassailable peaks here in Burgundy.

The taste, colour and aroma of a Burgundy Pinot Noir varies according to its village of origin (even according to each parcel of vines), but it is always marked by an evocative fruity flavour, balanced acidity and a signature touch of minerality. In fact, the name of the grape is never written on the label of a Burgundy wine, just the vineyard, known as a *climat*, along with the official classification of appellation – either 'Village', 'Premier Cru' or the ultimate accolade, a 'Grand Cru'.

Despite producing some of the world's most celebrated wines for over 2000 years, most Burgundy winemakers, whose families have often owned their vineyards for centuries, are down-to-earth and welcoming. This friendly reception is appreciated: tasting wines here can feel a little intimidating, due to both the incredibly high quality and the incredibly high prices. But once you sit down at a rough wooden table in a rustic cellar with a cheerful, ruddy *vigneron*, gently swirling a glass of subtly coloured Pommard or golden-hued Montrachet, it is impossible not to succumb to the Burgundy charm. It's a million miles from the glitzy world of a three-star Michelin restaurant where the same vintage is being carefully poured by a smartly dressed sommelier.

GET THERE

Geneva is the nearest major airport, 269km (167 miles) from Fixin. Car hire is available. The train from Paris to Dijon takes 1hr 33min.

DIJON
01
02
VOSNE-ROMANÉE
NUITS-ST-GEORGES
03
04
BEAUNE
POMMARD
05
06
07
FRANCE

01 DOMAINE JOLIET PÈRE ET FILS

Fixin, pronounced 'Fissin', is just outside Dijon at the beginning of La Route des Grands Crus. As the sleepy village does not actually boast a Grand Cru, it's often bypassed as enthusiasts speed on to the mythical vineyards of Chambolle and Vougeot. But Bénigne Joliet makes exceptional red wines, and his tiny domaine is unchanged since the day it was built in 1142 by Benedictine monks. Sitting in the immense vaulted wine cellar, complete with a medieval wooden press, Bénigne explains how this magical place inspires him. 'Every morning when I come in to check the barrels I imagine the scene a thousand years ago: the monks dressed in their habits, going out to work in the vines, no phones, no computers. I feel privileged to make wine here.'

www.domainejoliet.fr; tel 03 80 52 47 85; Manoir de la Perrière, Fixin; Mon–Sat by appointment

02 DOMAINE RION

Vosne-Romanée looks like just another tranquil Burgundy village. But behind almost every anonymous gate lie some of the world's most famous vineyards: Romanée-Conti, La Tâche, Richebourg. Dating back to at least the 11th century, these wines are, quite simply, priceless. Not surprisingly, few of these famous names are open for visits, but the Rion family can claim a long heritage as residents, even if today their cellars are on the busy road linking Beaune with Dijon. 'My father decided back in the 1950s that it was just not practical to be based in the village', explains fifth-generation *vigneronne*, Armelle Rion. 'Small producers like us had to wash our barrels out on the pavement, and delivering the grapes during harvest was a nightmare. Here we have plenty of room to both make and age our wine, as well as receiving visitors to taste our Vosne Romanée, Clos de Vougeot and Chambolle Musigny.'

www.domainerion.fr; tel 03 80 61 05 31; 8 Route Nationale, Vosne-Romanée; Mon–Sat

03 DOMAINE CAPITAIN-GAGNEROT

One of the oddities of Burgundy is that with so many domaines owning

01 Vergisson

02 Pommard

03 Domaine Rion wines

04 Winemakers Joannes and Thierry Violot-Guillemard

05 Burgundy's patchwork of vineyards

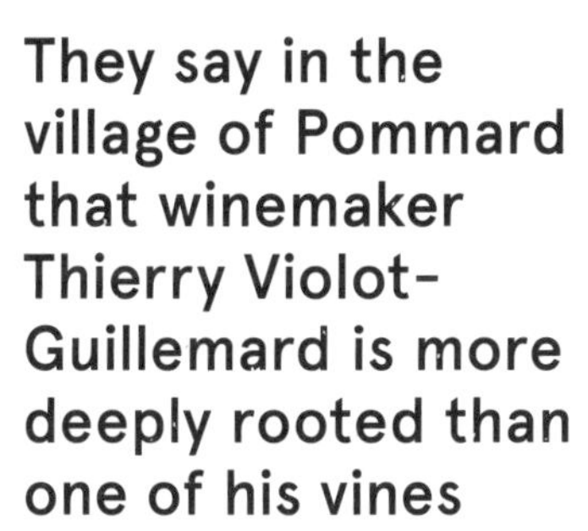

They say in the village of Pommard that winemaker Thierry Violot-Guillemard is more deeply rooted than one of his vines

small parcels, it is by no means assured that the best wines come from a cellar actually in the village that bears the name of a Grand Cru. This is certainly the case for Aloxe-Corton, an utterly idyllic hamlet with a fairy-tale castle. Few tourists stop off in busy adjacent Ladoix, and the Domaine Capitain-Gagnerot sits on the trunk road to Dijon. But visitors are warmly welcomed to try the excellent selection of Corton Grands Crus, including an outstanding white Corton-Charlemagne, as well as the less-renowned but complex Premier Cru of Ladoix itself. *www.capitain-gagnerot.com; tel 03 80 26 41 36; 38 Route de Dijon, Ladoix-Serrigny; daily by appointment*

04 THIERRY VIOLOT-GUILLEMARD

They say in the village of Pommard that Thierry Violot-Guillemard is more deeply rooted here than one of his vines. Certainly, a tasting in his tiny cellar, or a stay in the family's B&B, is the perfect introduction to the ingrained hospitality of a typical Burgundy *vigneron*. His 6-hectare (15-acre) organic estate spreads over Volnay, Beaune, Monthélie and Meursault, but Thierry is defined by his incomparable interpretation of Pinot Noir in Pommard. 'Our most important work is always in the vineyard, with as little intervention in the cellar as possible,' he philosophises, 'and then, the wine must be left alone to age.' *www.violot-guillemard.fr; tel 03 80 22 49 98; 7 Rue Sainte-Marguerite, Pommard; Mon–Sat by appointment*

05 DOMAINE GLANTENAY PIERRE ET FILS

With its commanding church and solemn war memorial, Volnay seems an austere village. But there

Ø6 Hospices de Beaune

Ø7 Escargots de Bourgogne – Burgundy snails

Ø8 Boating on the Burgundy canal

is nothing austere about the elegant wines presented in the homely tasting room of the Glantenay family. Pierre Glantenay has handed the reins to his 27-year-old son, Guillaume, who has built a dazzling modern cellar. 'I am committing myself to the domaine for a very long time with this kind of financial investment,' he says, 'but we have such a unique terroir that with the new cellar I really believe I can make even more exceptional wines than my father.' *www.domaineglantenay.com; tel 03 80 21 61 82; 3 Rue de la Barre, Volnay; Tue–Sat by appointment*

06 DOMAINE YVES BOYER MARTENOT

With its grand château-like town hall covered with zigzagging coloured roof tiles, Meursault is one of Burgundy's famed destinations. Vincent Boyer is a down-to-earth winemaker, producing a selection of distinctively mineral Premier Crus: 'Our vines are mostly very old, many over 90 years, which means low yields but high quality. I have been moving towards organic for the last few years, all but eliminating the use of chemicals, but I won't yet sign up for the inflexibility of certification.' *www.boyer-martenot.com; tel 03 80 21 26 25; 17 Place de L'Europe, Meursault; Tue–Sat*

07 DOMAINE JEAN CHARTRON

The most famous name in Burgundy for white wine is Montrachet. Five Grands Crus are concentrated around Chassagne-Montrachet and Puligny-Montrachet villages. Visiting the modern winery of Jean-Michel Chartron, he dips the tasting pipette into a barrel and describes what it's like to make what is often known as the world's greatest wine: 'I don't feel that I own my vines, but rather that I am a guardian for the centuries of workers who have toiled the land before and the future generations. My name on the bottle is not important.' After wandering through these historic vineyards, most people can't resist buying at least one bottle, regardless of the expense. A 10-year-old Bâtard-Montrachet, say, with its intense aromas and flavours of apple, almonds and spices, is perfect to savour over a romantic meal at home, accompanying perhaps scallops or lobster. *www.jeanchartron.com; Grand Rue, Puligny-Montrachet; tel 03 80 21 99 19; Thu–Sat*

ESSENTIAL INFORMATION

WHERE TO STAY

CHÂTEAU DE GILLY
A fairy-tale castle near the iconic Clos de Vougeot and vineyards of Romanée-Conti. Four-poster beds and a 14th-century vaulted dining room await. *www.chateau-gilly.com; tel 03 80 62 89 98; Gilly-les-Citeaux, Vougeot*

CHAMBRES D'HÔTES DE L'ORMERALE
The winemaking Fouquerand family run a simple cottage B&B in this delightful village. Be sure to sample the local bubbly, Crémant de Bourgogne. *www.domaine-denisfouquerand.com; tel 03 80 21 88 62; Rue de l'Orme, La Rochepot*

CHÂTEAU DE MELIN
A romantic château B&B with a verdant park and lake. Each evening in the medieval cellar Arnaud Derats hosts a tasting of his wines, which originate from small parcels from Meursault to Chambolle-Musigny. *www.chateaudemelin.com; tel 03 80 21 21 29; Hameau de Melin, Auxey-Duresses*

WHERE TO EAT

LA TABLE DE PIERRE BOURÉE
Winemaker Bernard Vallet has created the perfect casual setting for his traditional Burgundy cuisine paired with top vintages. Perhaps you could try *blanquette de veau* with a 2016 Gevrey-Chambertin. *www.pierre-bouree-fils.com; tel 03 80 34 13 97; 40 Route de Beaune, Gevrey-Chambertin*

LA TABLE D'OLIVIER
Respected *vigneron* Olivier Leflaive has transformed Puligny-Montrachet with his hotel and restaurant, its menu tailored for wine pairings. *www.olivier-leflaive.com; tel 03 80 21 37 65; Place du Monument, Puligny-Montrachet*

BOISROUGE
Feast on roast suckling pig with crunchy cabbage at chef Philippe Delacourcelle's gourmet venue, where you can also stay the night or sign up for a cookery course. *www.boisrouge.fr; tel 03 80 34 30 56; 4bis Rue du Petit Paris, Flagey-Echézeaux*

WHAT TO DO

Intermingle boating, cycling and wine tasting courtesy of a barge cruise along Burgundy's historic canal. *www.burgundy-canal.com*

CELEBRATIONS

On the third Sunday in November, the Hospices de Beaune holds its world-famous charity wine auction, as it has done since 1851.

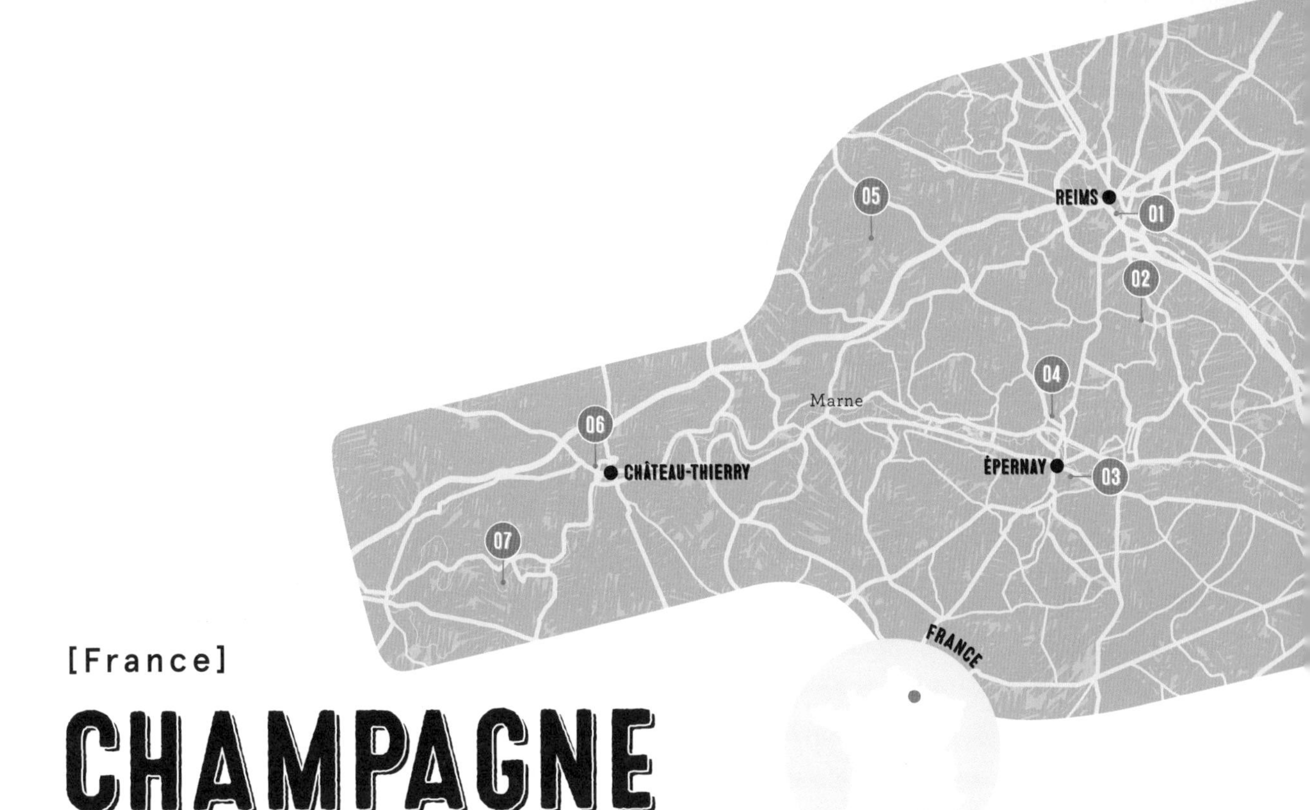

[France]

CHAMPAGNE

Pop! The land that produces the king of sparkling wines is a treasure trove of rolling hills, ancient cellars and traditions just waiting to be opened.

Champagne is France's great enigma: the world's most famous bubbly and an undisputed icon of Gallic glamour, yet most French people have little idea of the complex, almost mystical, secrets that go into producing Champagne.

A trip into this magical land, on Unesco's World Heritage list since 2015, is an emotional experience. You can witness perfectly cultivated vines hanging with juicy clusters of grapes that will soon begin the long transformation into the one and only Champagne. Take a pilgrimage through the centuries-old maze of cellars beneath the likes of Ruinart or Pommery that resemble a holy subterranean cathedral, or savour the simple pleasure of a smallholder *vigneron* pouring a bubbly glass of his latest vintage. Accept the sly persuasion that he may sell most of his grapes to the famous producers, but keeps the best for personal production sold directly from the independent winery.

The region's bucolic vineyards stretch across rolling hills and sleepy villages that begin just an hour's drive from Paris, although wine-lovers often limit themselves to a trip to Reims. A regal city, Reims is home to the likes of Veuve Clicquot and Mumm, where incredible cellars, storing millions of bottles, are packed every day for tours. But Champagne is a complex mosaic of thousands of tiny *vignerons*, some making their own Champagne, others just supplying grapes to the luxury Champagne houses, an almost feudal relationship unchanged for centuries. So, after visiting Reims, meet these independent winemakers, who will explain the blending of Champagne's three grapes – Chardonnay, Pinot Noir and Meunier – the difference between *millésime* vintage and an NV (the anglicised non-vintage), and insider secrets like the use of the liqueur de dosage of cane sugar added in the final *assemblage*.

GET THERE

Charles de Gaulle is the nearest airport, 130km (81 miles) from Reims. The train from Paris to Reims takes 50min. Car hire is available.

01 REIMS

Reims is home to Champagne's own royalty, with curious visitors allowed into the hallowed cellars of the likes of Mumm and Pommery, Veuve Clicquot, Heidseck, Lanson and Taittinger. It is the perfect place to get an idea of the hidden secrets of arguably the world's favourite beverage, with gushing guides explaining the centuries-old alchemy that goes into its production. Each of the *Grandes Maisons* offers something different, but which one to choose? Taittinger stands out as being one of the rare family-owned houses, and its two-level 13th-century cellars are primarily reserved for ageing the signature vintage Comtes de Champagne, a remarkable cuvée. The neo-Gothic castle towers of Pommery resemble a kitsch Disneyland, but this is the one must-see cellar. Madame Pommery, 140 years ago, conceived dry Brut Champagne as a counterpoint to sweet bubbly, and her 18km (11 miles) of cellars are like no others. This is where you will discover *les crayères*, some 120 awesome chalk pits dug beneath Reims in Gallo-Roman times. Madame Pommery decided these provided perfect ventilation for the maze of tunnels she built for her cellars, which today hold some 20 million bottles.
www.taittinger.com; www.vrankenpommery.com

02 CHAMPAGNE GARDET

Over two thirds of all Champagne, including 90% exported around the world, is produced by the 290 *Negociants Manipulants*, the *Grandes Maisons* who own hardly any vines but purchase grapes at harvest. Gardet, founded in 1895, is still a relatively small maison, owning a mere 5 hectares (12 acres) of vineyards, but produces a million bottles a year using grapes from another 100 hectares (247 acres) it buys in. Visitors are received in an ornate art nouveau glass veranda filled with tropical plants. A tour of the *cuverie*, where the wine is made, and the labyrinth of cellars, takes over an hour and gives a thorough explanation of all the stages of Champagne's complex production.

01 Champagne region vineyards

02 Hautvilliers, the birthplace of Champagne

03 Electric train at Mercier cellars

04 Pouring out the bubbly

www.champagne-gardet.com; tel 03 26 03 42 03; 13 Rue Georges Legros, Chigny-les-Roses; Mon–Sat by appointment $

03 CHAMPAGNE MERCIER

Lively Épernay is the genuine wine capital of Champagne, with a host of fun wine bars, gourmet restaurants and bistros. Over 100,000 visitors arrive each year at Mercier, one of the most popular Champagnes in France itself. Founder Eugene Mercier was the publicity-seeking Richard Branson of his time, building an immense wooden barrel holding 250,000 bottles of Champagne that was transported by oxen to Paris in 1900 to rival the Eiffel Tower as the star show of the Exposition Universelle. Today the barrel dominates the entrance of Mercier's outstanding cellar, where a lift plunges visitors into an eerie maze of tunnels. A small electric train wends part of the way, and you realise how deep underground the cellar workers are.

www.champagnemercier.fr; tel 03 26 51 22 22; 68 Ave de Champagne, Épernay; daily $

04 CHAMPAGNE TRIBAUT

Before arriving for a tasting at the friendly Tribaut family winery, take a tour of the idyllic village of Hautvilliers, known as the birthplace of Champagne. There is a Rue Dom Pérignon, named after the Benedictine monk who, 300 years ago, is said to have invented the process of double fermentation that creates Champagne's unique bubbles. Ghislain and Marie-José Tribaut, along with their daughter and grandson, love to welcome wine tourists. 'I am a Récoltant Manipulant,' explains Ghislain, 'someone who cultivates and harvests their grapes, and can then sell them to a Négociant Manipulant – Grandes Maisons like Krug and Taittinger. But personally, I keep them all for myself, enough to produce 200,000 bottles of our own Champagne.' After tasting Marie-José's delicious *gougères* (light puff pastry filled with Gruyère), paired with a dry Rosé Brut, many visitors end up coming back here to help out during the grape harvest.

05 Pannier's 12th-century cellars

06 Notre-Dame cathedral, Reims

www.champagne.g.tribaut.com; tel 03 26 59 40 57; 88 Rue d'Eguisheim, Hautvillers; daily $

05 CHAMPAGNE ASPASIE

Paul-Vincent Ariston is an artisan *vigneron* who bubbles with as much enthusiasm as his Champagne. Spend the night in his comfy B&B, allowing time for a cellar visit down below the 400-year-old stone farmhouse, where Paul-Vincent proudly shows a huge wooden grape press, ancient but functioning, then explains the *dégorgement*, when sediment is frozen in the neck of the bottle and spectacularly popped out before final bottling. Try the Brut de Fut, aged in oak barrels, while his unique cuvée, Brut Cépages d'Antan, has none of the usual Champagne grapes but, rather, three rare varieties – Le Petit Meslier, L'Arbanne and Pinot Blanc – that were grown here centuries before Champagne was popularised.

www.champagneaspasie.com; tel 03 26 97 43 46; 4 Grande Rue, Brouillet; Mon-Sat $

06 CHAMPAGNE PANNIER

Pannier is one of Champagne's better-known cooperatives, a Coopérative de Manipulation to use the official title. It features a breathtaking labyrinth of cellars – stretching 30m (33yd) beneath the earth – which date back to the 12th century when they were excavated to build churches. A small group of 11 *vignerons* formed the original cooperative in 1974, which has since mushroomed into a vast winery representing 400 growers. Although it produces millions of bottles a year, the cooperative keeps Pannier separate as its prestige brand, blending the local Meunier grape with Chardonnay and Pinot Noir from vineyards from the faraway Côte des Blancs and Montagne de Reims.

www.champagnepannier.com; tel 03 23 69 51 30; 23 Rue Roger Catillon, Château-Thierry; Mon-Sat $

07 CHAMPAGNE FALLET DART

Just an hour's drive from Paris, this part of the agricultural Marne valley was only incorporated into the exclusive members-only club of the Champagne appellation in 1937. Paul Dart is a dynamic young winemaker, and although the estate is medium-sized, stretching over 18 hectares (44 acres), it still has something like one million bottles ageing in its cellar. Be sure to taste the Clos du Mont, a blend of vintages from a vineyard dating from the 7th century. Dart is also proud of the domaine's Ratafia, a luscious aperitif, and an elegant Fine de Champagne, aged in barrels like a Cognac.

www.champagne-fallet-dart.fr; tel 03 23 82 01 73; 2 Rue des Clos du Mont, Drachy, Charly sur Marne; Mon-Sat

ESSENTIAL INFORMATION

WHERE TO STAY

LES CRAYÈRES
Built by the family of Madame Pommery, this grand château. sumptuously furnished, with a two-star Michelin restaurant, offers the full Champagne experience. *www.lescrayeres.com; tel 03 26 24 90 00; 64 Blvd Henry Vasnier, Reims*

PARVA DOMUS
Claude and Ginette Rimaire pamper guests in their cosy home on Avenue de Champagne, which Churchill named 'the world's most drinkable address'. Hearty breakfasts and a glass of Champagne on arrival are included. *www.parvadomusrimaire.com; tel 06 73 25 66 00; 27 Avenue de Champagne, Épernay*

WHERE TO EAT

LA GRILLADE GOURMANDE
A favourite restaurant where *vignerons* rub shoulders with owners of the *Grandes Maisons*. Try the hearth-grilled meat or delicate dishes such as pigeon stuffed with foie gras. *www.lagrilladegourmande.com; tel 03 26 55 44 22; 16 Rue de Reims, Épernay*

06

AU 36
A perfect spot for food and Champagne pairing, this designer bar serves plates of local specialities – creamy Chaource cheese, Reims ham, smoky lentils and pink macarons – served with three different Champagnes. *www.au36.net; tel 03 26 51 58 37; 36 Rue Dom Pérignon, Hautvillers*

BISTROT LA MADELON
Far from Champagne's many gourmet dining rooms, this old-fashioned village bistro serves a generous plat du jour, such as slow-cooked veal. *Tel 03 26 53 14 18; 7 Grande Rue, Mancy*

WHAT TO DO

Notre-Dame de Reims is a must-see 800-year-old Gothic cathedral, and the historic venue for the coronation of the kings of France. Its interior is adorned with intricate stained-glass windows.

CELEBRATIONS

Épernay celebrates Habits de Lumière for three days in mid-December, when Champagne flows amid fireworks, flamboyant light shows and street theatre. *https://habitsdelumiere.epernay.fr*

01

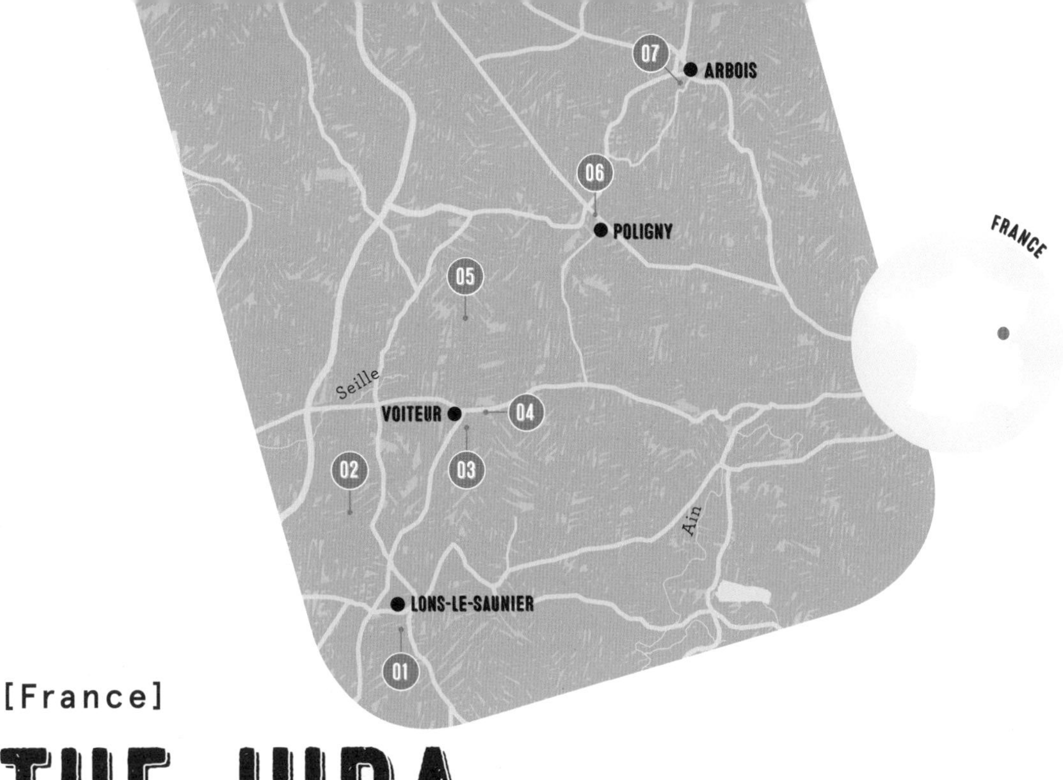

[France]

THE JURA

This oft-overlooked region tucked away on the Swiss border, quietly making wine for centuries, holds some quirky surprises for even experienced wine-tourers.

Wine has been made in the mountainous Jura for well over a thousand years. But it is only recently that this corner of France has begun to make a name for itself in the world of wine. The region has a rich biodiversity of lush valleys and thick forests, with vineyards adjacent to agricultural and grazing land. The majority of grapes cultivated are little-known indigenous varieties, from delicate light reds – Trousseau, Poulsard – to the remarkable Savagnin, which makes a white wine perfect for long ageing. A new generation of *vignerons* are making their mark here, using modern winemaking techniques alongside the Jura's traditional method.

And nothing quite prepares you for a tasting of the extraordinary Vin Jaune. No one is left undecided about Vin Jaune, so be prepared to love it or hate it. While the distinctive aroma immediately seems different from other wines – a mix of walnuts, hazelnut and exotic spices – the taste is altogether something else, incredibly dry yet somehow fruity and nutty at the same time. The wine is a brilliant cooking ingredient, with all Jura households stocking a bottle to add to dishes such as chicken with morel mushrooms. It's also an ideal pairing with the Jura's tart Comté cheese. Made purely from Savagnin grapes, Vin Jaune is barrel-aged for six years, but with a pocket of air left open; its oxidising effect is limited as the maturing wine is covered by a natural *voile*, a film of yeast, in the same way Spanish sherry is produced in Jerez.

Wine tourism is still in its early days in the Jura, but that makes for an even more refreshing welcome when travellers turn up to taste in a little-known backwoods domaine. For now, these young Jura *vignerons* are concentrating all their efforts on winemaking, though many already have plans to convert parts of their rambling stone farms into holiday homes.

GET THERE
Geneva is the nearest major airport, 143km (89 miles) from Montaigu. Car hire is available.

01 DOMAINE PIGNIER

A tasting at this historic domaine is the perfect introduction to the Jura. Ask Marie-Florence Pignier, a seventh-generation *vigneronne*, to take you down for a tour of the astonishing 13th-century cellar. The vast, high, vaulted barrel-chamber resembles a cathedral, so it's not surprising to learn that this was formerly a monastery, founded in 1250 by *vigneron*-monks who planted the original vineyard. The wines produced today are organic and highly contemporary, and certain cuvées are *naturel*, with no sulphite added. Don't miss the Vin Jaune either, which Marie-Florence proudly claims 'is the ultimate wine for ageing, even if you want to wait a century!' *www.domaine-pignier.com; tel 03 84 24 24 30; 11 Place Rouget de Lisle, Montaigu; Mon-Sat*

02 DOMAINE VANDELLE

Étoile sits in a bucolic valley encircled by a series of rolling hills covered with vineyards and woods. 'Pretty much each hill and its vines is owned by a different village *vigneron*,' recounts Philippe Vandelle, 'and each of us make wines with a different personality due to the variations of soil and exposure to the sun.' The Vandelles came to the Jura over two centuries ago from Belgium, and a cousin still owns the grand Château de l'Étoile winery. Philippe has converted a stone labourer's cottage into a snug tasting room, and it is a surprise to learn that 30% of his production is devoted to Crémant du Jura, made following the classic Champenoise method. *www.vinsphilippevandelle.com; tel 03 84 86 49 57; 186 Rue Bouillod, L'Étoile; Mon-Sat*

03 FRUITIÈRE DE VOITEUR

As you drive out of Voiteur, you can't miss the massive and spectacular limestone outcrop with Château-Chalon balancing on its summit to the left of the road, while opposite is an imposing modern winery. 'A Jura *fruitière* has nothing to do with fruits,' explains Bertrand Delannay, the director here, 'but is rather an agricultural cooperative devoted to one of the region's

01 Château-Chalon

02 Village views from Château-Chalon

03 European bee-eaters in the Jura

two specialities – wine or cheese.' There are a challenging 19 wines to taste, with eminently affordable prices and a lot of variety. 'Many Jura *vignerons* specialise only in barrel-aged wines as that is the tradition here, but we try to offer some easier-to-drink alternatives too, such as a young floral Chardonnay aged in steel vats.' *www.fruitiere-vinicole-voiteur.fr; tel 03 84 85 21 29; 60 Rue de Nevy, Voiteur; daily*

04 DOMAINE CREDOZ

Medieval Château-Chalon, classified as one of France's most beautiful villages, looks down on a criss-cross patchwork of vineyards, including 9 hectares (22 acres) cultivated by Jean-Claude Credoz, an innovative *viticulteur* (winemaker). His excellent sparkling Crémant is sold out a year in advance, while enthusiasts come just to taste his Macvin, the unique Jura aperitif that he makes from the must (juice) of Savagnin grapes and a distilled marc (pulped skins) aged for four years in oak. Working essentially old vines, some over 80 years, and ageing from three to seven years in ancient barrels, his white wines – Chardonnay, Savagnin and Vin Jaune – are elegant and subtle on initial tasting, but with all the grape's delicate expression coming out in what Jean-Claude lovingly describes as the *longueur* of the aftertaste.
Tel 03 84 44 64 91; 3 Rue des Chèvres, Château-Chalon; Mon–Thu by appointment

05 LES DOLOMIES

Céline Gormally only founded her domaine in 2008. 'There is a great feeling of solidarity here,' she explains. 'I immediately sought organic certification... and when I first started, I was able to rent parcels of wonderful 70-year-old vines at a fair-trade price from an agricultural association.' To lessen the financial burden, Céline has started her own private club, Location de Cep, where members order wine for the forthcoming vintage but pay a year in advance, with many coming as unpaid help during the harvest. Her whites are bottled by individual parcels of vines, and the Pinot Noir is surprisingly full-bodied for the Jura. And ask to try naturally fermented Tout Pet bubbly.

www.les-dolomies.com; tel 06 87 03 39 98; 40 Rue de l'Asile, Passenans; daily by appointment

06 DOMAINE BADOZ

Bernard Badoz launched the Percée du Vin Jaune festival in 1977, where every *vigneron* in the Jura presents their wines, and which launched this little-known region on to the world wine map. Visitors arrive at this modern boutique where they can taste Comté cheeses from a cousin's farm, organic honey and regional artisan charcuterie such as smoky Morteau sausage. Bernard's son, Benoit, who runs the organic estate today, has created a new range of special cuvées; 'Edouard' is Chardonnay aged in barrels made especially from wood in the forest above Poligny, while 'Arrogance', which he modestly named for himself, sees a crisp, acidic Savagnin aged normally rather than oxidised with the traditional *voile* of yeast used for Vin Jaune.

www.domaine-badoz.fr; tel 03 84 37 18 00; 19 Place des Déportés, Poligny; daily

07 DOMAINE RIJCKAERT

Arbois is the lively winemaking capital of the Jura, but the winery of dynamic *vigneron* Florent Rouve is hidden just outside town in a sleepy hamlet. Florent is always renovating and innovating, so there is no sign outside his ancient farmhouse and no proper tasting room, but that doesn't detract from the joy of discovering his artisan vintages. He is a white-wine fanatic, working exclusively with Chardonnay and Savagnin, so don't expect any reds. Informal but animated tastings take place down in a 17th-century vaulted cellar that is mouldy, cold and humid – perfect conditions, according to Florent, to age using the traditional *voile* method. 'I press the juice, bring it down into the barrels and begin ageing on the lie. Then wait. It really isn't complicated to make good wine, you just need patience,' he says with a wry smile.

www.vinsrijckaert.com; tel 06 21 01 27 41; Villette-les-Arbois; daily by appointment

04 Domaine Credoz

05 Salins-les-Bains

ESSENTIAL INFORMATION

WHERE TO STAY

LE RELAIS DE LA PERLE
Nathalie Estavoyer welcomes travellers to her beautifully restored *maison de vigneron*, organising wine tastings and even a hot-air balloon trip high above the vineyards.
www.lerelaisdelaperle.fr; tel 03 84 25 95 52; 184 Route de Voiteur, Le Vernois

LE DORTOIR DES MOINES
The magnificent Romanesque abbey of Baume-les-Messieurs is already one of the Jura's most spectacular sights. Stay in one of its private apartments, now transformed into magical accommodation.
www.dortoir-des-moines.info; tel 06 33 21 21 46; L'Abbaye, Baume-les-Messieurs

WHERE TO EAT

BISTROT CHEZ JANINE
Madame André keeps raucous *vignerons* quiet with her hearty *planche comtoise*, heaped with cheeses, smoked ham, home-pickled gherkins and *saucisson*.
Tel 03 84 44 62 43; Route de la Vallée, Nevy-sur-Seille

LA FINETTE TAVERNE D'ARBOIS
A rustic wooden chalet in Jura's winemaking capital. Feast on regional specialities such as succulent slow-cooked chicken in Vin Jaune.
www.finette.fr; tel 03 84 66 06 78; 22 Avenue Louis Pasteur, Arbois

LE PONTARLIER
With its red-checked tablecloths and zinc bar, this classic bistrot oozes Gallic charm, and chef Julien Zangiacomi prepares traditional dishes including local-cheese platters and frogs' legs in a garlicky parsley sauce.
www.bistrotdeportlesney.com; tel 03 84 37 83 27; Place du 8 Mai 1945, Port-Lesney

WHAT TO DO

Visit the subterranean salt mines of the Unesco World Heritage-listed Salins-les-Bains.
www.salinesdesalins.com

CELEBRATIONS

Every year, on the first weekend of February, a different village hosts La Percée du Vin Jaune, toasting the new vintage.
www.percee-du-vin-jaune.com

05

[France]

THE LANGUEDOC

Craggy cliffs and wooded valleys greet visitors to this dynamic wine region, fast becoming one of France's most exciting.

Some of the most interesting and innovative wines in France are emerging from the vast Languedoc-Roussillon region, which covers much of the south, from the Spanish border up to the vineyards of Provence and the Côte d'Azur. A third of France's wine is produced here, but for years the region suffered from poor quality and over-production. Not any more. The winery scene has changed dramatically, with a flood of new appellations, the huge popularity of Vin de Pays wines, and advances in both vineyard and cellar. Dynamic young *vignerons* are drawn here, not to contribute to the old system of the winemaking cooperative, but to set up their own small vineyards, often organic and biodynamic. There are new regional stars: intense reds from Pic St-Loup and La Clape, where many *vignerons* are experimenting with ageing wines in terracotta jars, as in Roman times; the bubbly Blanquette de Limoux; and crisp white Picpoul de Pinet, perfect with local oysters. And one patch, the Corbières, wedged between Montpellier and Perpignan, remains under the radar, waiting to be discovered.

GET THERE
Toulouse is the nearest major airport. The train from Paris to Narbonne takes 4hrs. Car hire is available.

The sheer variety of the landscapes here is spectacular, with vineyards pressing up against the Mediterranean along the flamingo-filled lagoons of Peyriac-de-Mer, through dramatic limestone hills and valleys where travellers can stay in *chambres d'hôtes* in isolated medieval hamlets, right up to the wild mountain castles built by the Cathar tribes in the 12th century in the foothills of the Pyrenees. Locating a restaurant in this rugged corner is not always easy, but when you do, you'll find the Grenache, Syrah and Carignan reds from Corbières are sturdy, spicy and robust, and perfect with the region's hearty cuisine.

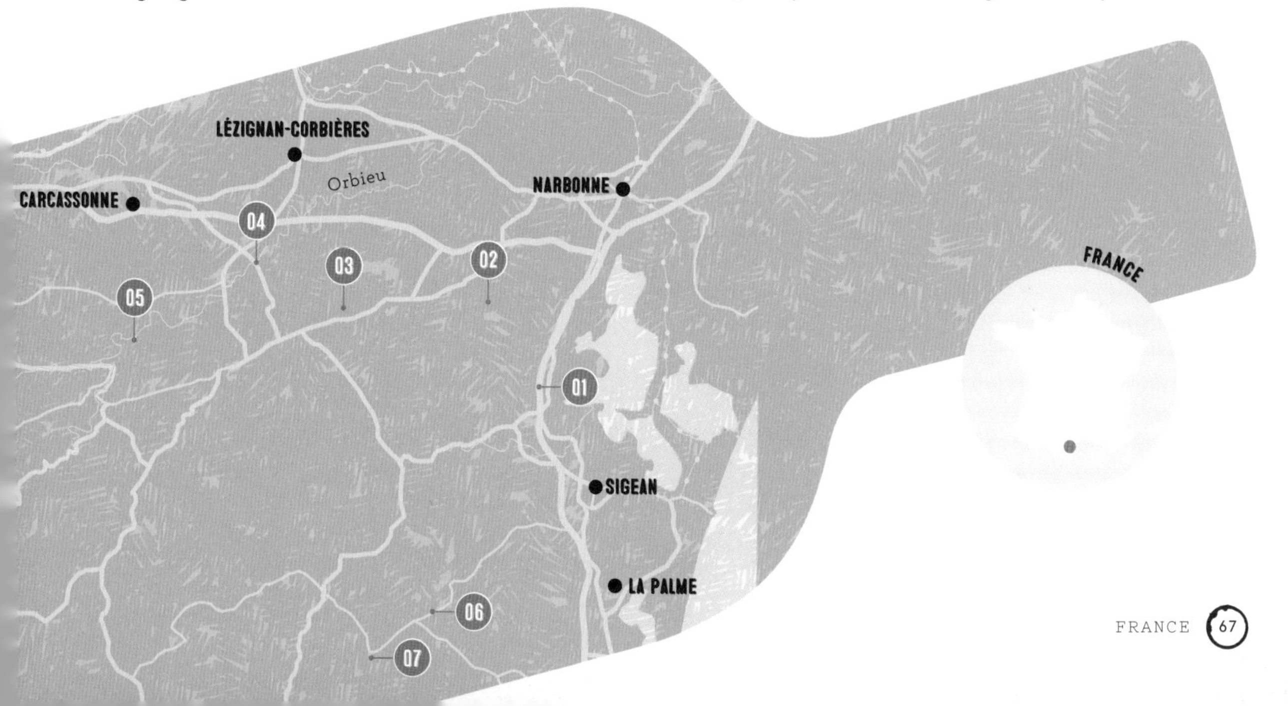

01 CHÂTEAU FABRE-CORDON

Amandine Fabre-Cordon is typical of the numerous passionate young women earning respect as independent *vigneronnes* in the Corbières today. She took over her father's estate in 2011, after learning her craft in New Zealand and California. The terroir here is known as Corbières Méditerannée, as the sea is just 3km (1.9 miles) away at the picturesque fishing village of Peyriac. She makes an especially strong selection of white and rosé organic wines – Grenache Blanc, Viognier and Vermentino. Other dynamic female winemakers to check out nearby are Cécile Bonnafous (www.domaine-esperou.fr) and Fanny Tisseyre (www.graindefanny.com).

www.chateaufabrecordon.fr; tel 06 87 84 15 46; L'Oustal Nau, Peyriac-de-Mer; Mon–Sat by appointment

02 ABBAYE DE FONTFROIDE

This magnificent medieval abbey is the perfect place to get a feel for the history and winemaking heritage of the Corbières. Fontfroide was founded in 1093 by monks, who immediately planted vines to provide wine for religious services. At its peak, these Cistercians controlled thousands of hectares. Abandoned at the end of the 19th century, the abbey was bought by the Fayet family in 1908, and although 40% of production goes to the local cooperative, they are producing some fine wines in their nearby modern cellar, especially a dry Muscat and Clôture, an elegant blend of Syrah and Grenache. The tour round the abbey and its wonderful gardens followed by a tasting is simply unforgettable.

www.fontfroide.com; tel 04 68 45 11 08; Route Départemantale 613, Narbonne; daily

03 CHÂTEAU LES OLLIEUX ROMANIS

As its name implies, this vast estate in the heart of the Boutenac Cru (a tiny but high-quality appellation within Corbières) has roots going back to Roman times, and today with 150 hectares (370 acres) of

01 Abbaye de Fontfroide

02 Fontfroide cloisters

03 Languedoc rosé

04 Laure d'Andoque of Abbaye de Fontfroide

Laurent and Sylvie Bachevillier are typical of the new generation of young vignerons choosing the Corbières as the ideal place to set up a winery

vines the château is one of the largest private domaines. But there is a very warm, human welcome provided by the enthusiastic team of *vignerons* that surround the owner, Pierre Bories. This is a still old-fashioned farm, with donkeys wandering about, sheep and goats grazing, and chickens running around everywhere, while Pierre is always accompanied by his faithful shaggy dog, Nounours (Teddy Bear). Pierre's parents made the crucial decision not to pull up and plant new vines in the 1980s, meaning he inherited a tremendous selection of vines, some 120 years old, growing in a mix of red clay and sandstone.
www.chateaulesollieux.com; tel 04 68 43 35 20; Route Départementale 613, Montséret; daily

04 DOMAINE LEDOGAR

The Ledogar family have been making wine in Ferrals for many generations, closely associated with the local winemaking cooperative until the arrival of uncompromising brothers Xavier, Mathieu and Benoit. They are what could be termed 'natural wine' fundamentalists, and since 1997 have created a sprawling 22-hectare (54-acre) vineyard, producing wines that are 100% organic, working the vine around a lunar calendar, with no sulphur and a hand-picked harvest. More than a dozen different grape varieties are cultivated – not just classic Carignan and Grenache but Mourvèdre, Maccabeu and Merselan too. There is a small tasting room in the centre of Ferrals, where discussions can get passionate.
Tel 06 81 06 14 51; Place de la République, Ferrals-les-Corbières; daily by appointment

05 Château Les Ollieux Romanis, bottled

06 Château de Peyrepertuse

05 DOMAINE LES CASCADES

Laurent and Sylvie Bachevillier are typical of the new generation of young *vignerons* choosing the Corbières as the ideal place to set up a winery, and have also opened a charming three-room B&B and eco-*gîte* adjacent to their cellar. The domaine revolves around the concept of biodiversity, producing not only organic wine but vegetables, saffron, truffles and olive oil too. Instead of using chemical insecticides, Laurent takes out their two donkeys and three fearsome Hungarian sheep to graze among the vines. The couple's wines will take you by surprise, especially Cuvée S, a natural wine, 100% Grenache with no sulphite added.

www.domainelescascades.fr; tel 06 88 21 84 99; 4bis Avenue des Corbières, Ribaute; daily by appointment

06 DOMAINE SAINTE-CROIX

The landscape changes as you climb into the dramatic jagged mountains of Hautes Corbières. There is very little agriculture, villages are few and far between, and ancient vines grow in a patchwork of small parcels on a variety of different soils – limestone, clay, schist, volcanic. It was this diversity that attracted adventurous English winemaker Jon Bowen and his wife, Elizabeth, to settle here 15 years ago. In the cellar, Jon works primarily with steel tanks, occasionally using old barrels to age, 'just to give an idea of the wood, nothing more'. These organic wines immediately have a strong identity, especially *mono-cépage* cuvées using ancient indigenous grapes like Aramon, Terret Gris and Grenache Noir. This is very much an anarchic, garage winery, with Jon setting up bottles and glasses on an old wooden barrel as a tasting table.

www.saintecroixvins.com; tel 06 85 67 63 88; 7 Avenue des Corbières, Fraïssé-des-Corbières; daily by appointment

07 CASTELMAURE

This historic winemaking cooperative dominates an isolated hamlet of 150 souls, lost in the wild, windswept Cathar mountains. Intrepid visitors receive a fantastic welcome in the surprisingly modern tasting room and sample some extraordinary wines. You can't escape the culture of the winemaking cooperative in the Corbières, a system that historically has given *vignerons* financial security but has hardly garnered a reputation for quality wines. Not here in Castelmaure, though. Founded in 1921, the cooperative has 62 participants, all characters, but none more so than Patrick Marien, president for 34 years. Antoine Robert, the new young winemaker, constantly experiments – ageing in the bottle rather than barrels, using the old method of cement vats, keeping dosages of sulphite low and devising strikingly creative labels. Each wine is a surprise, from the uncomplicated La Buvette, 'our *vin de soif*' (easy to drink when you're thirsty), to the barrel-aged N°3 Corbières.

www.castelmaure.com; tel 04 68 45 91 83; 4 Route des Canelles, Embres-et-Castelmaure; daily

ESSENTIAL INFORMATION

WHERE TO STAY

CHÂTEAU DE L'HORTE
A winemaker B&B set in a grandiose 18th-century château with the four bedrooms located over the vast *chai* (barrel room). There's a pool along with a garden terrace for barbecues, plus tours and tasting on-site.
www.chateaudelhorte.fr; tel 04 68 43 91 70; Route d'Escales, Montbrun-des-Corbières

CHÂTEAU DE LASTOURS
Lastours combines a state-of-the-art winery, giant outdoor contemporary sculptures, a restaurant and 10 B&B rooms in discrete cottages.
www.chateaudelastours.com; tel 04 68 48 64 74; Portel-des-Corbières

WHERE TO EAT

O VIEUX TONNEAUX
Peyriac is famous for its wetland lagoons and flamingos, and for this cosy bistro. Cristelle Bernabeu cooks a tempting *bourride d'anguille* (eel stew).
Tel 04 68 48 39 54; 3 Place de la Mairie, Peyriac-de-Mer

PLACE DU MARCHÉ
Rub shoulders with *vignerons* in this lively gourmet bistro where Eric Delalande serves the likes of duck magret grilled with wild garrigue herbs.
www.placedesmarches-restaurant.com; tel 04 68 70 09 13; 8 Avenue de la Mairie, Villesèque-des-Corbières

CHEZ BEBELLE
Narbonne's historic covered market heaves at lunchtime as crowds teem around the stall of ex-rugby star Gilles Belzons, who theatrically shouts orders over a megaphone while grilling his delicious steaks and sausages.
www.chez-bebelle.fr; tel 06 85 40 09 01; Halles de Narbonne, 1 Blvd Dr Ferroul, Narbonne

WHAT TO DO

The Corbières mountains are marked by awesome clifftop Cathar castles, dating from a religious 12th-century war. Follow the route of the châteaux – and don't miss the precariously situated Château de Peyrepertuse.
www.payscathare.org

CELEBRATIONS

The small town of Conhilac gets taken over for a month-long jazz festival each November.

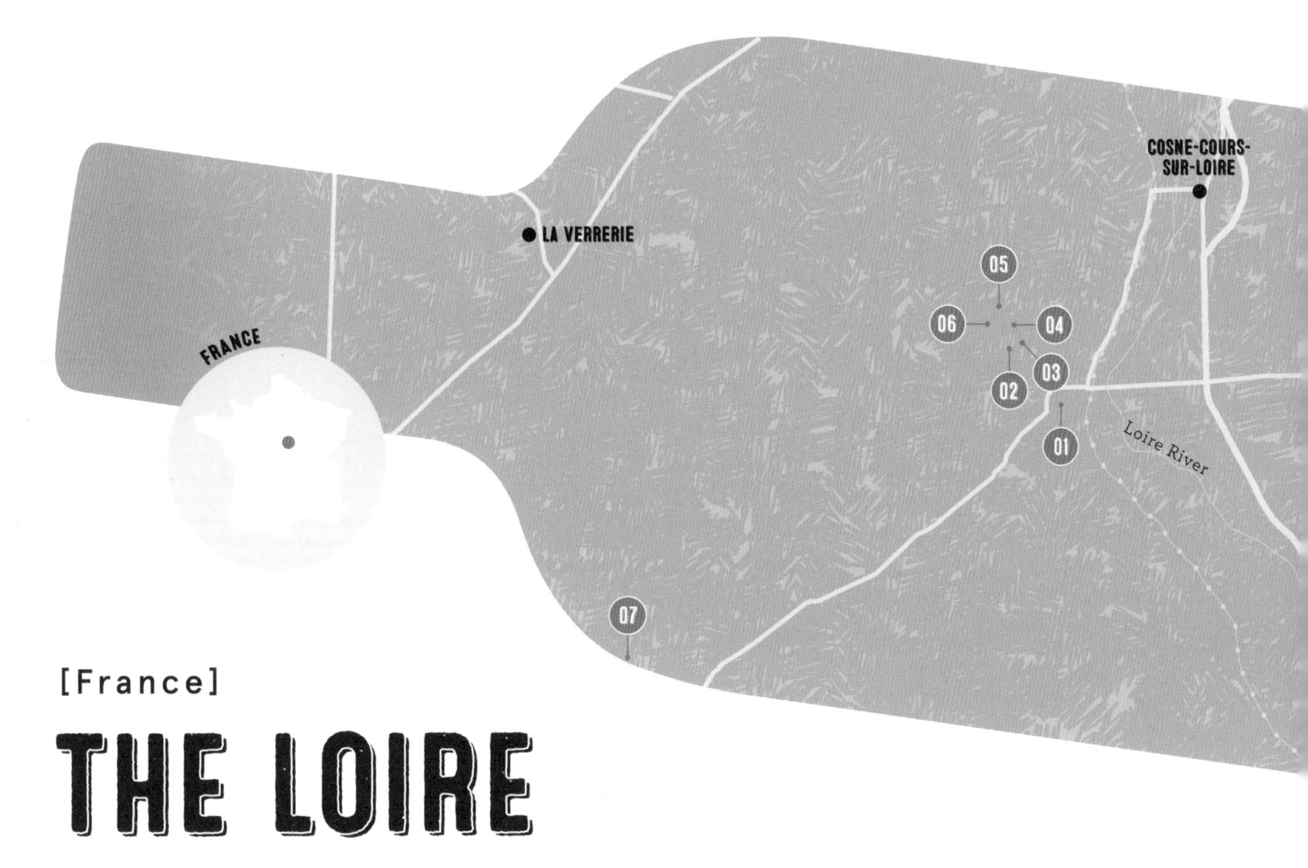

[France]

THE LOIRE

Take your time and explore the crisp white wines and fairy-tale châteaux of the languid Loire River in central France by boat, bicycle or car.

The Loire is the longest river in France, and along its banks some of the nation's most famous and varied wines are cultivated: the sharp white Muscadet and Anjou, sparkling Vouvray, and the fresh tannins expressed in the Cabernet Franc grape of Chinon and Saumur. And just 200km (124 miles) from Paris are the remarkable vineyards of Sancerre, from which distinctive Sauvignon has become one of the world's best-loved wines. The grand medieval town of Sancerre sits majestically atop a vine-clad hill overlooking the dozen villages that cover the appellation. The first reference to wines from Sancerre date back to 583 AD when Gregory of Tours mentioned the vintages here. Up until the phylloxera epidemic of 1886, the majority of production was actually red wine – Pinot Noir – and it was only when the vines were replanted that the decision was made to change to the now famous Sauvignon Blanc.

Few people who today tour the state-of-the-art cellars of wealthy winemakers realise that just one generation ago there was a great deal of poverty here, with *vignerons* struggling to sell a little-known and little-respected wine, their families only surviving thanks to the financial contribution of their wives, who raised goats to make Crottin de Chevignol cheeses. Sancerre owes its present fame and success to the young generation of vintners who took their wines up to Paris after the end of WWII, opening wine bars, convincing the capital that Sancerre was a fashionable wine and, at the same time, becoming early exponents of wine tourism by encouraging Parisians to visit Sancerre, see the vineyards and buy direct from the winemaker. Wine tourists still get a very special welcome – and now the whole of the world seems to have fallen in love with Sancerre.

GET THERE
Paris Charles de Gaulle is the nearest major airport, 226km (140 miles) from Sancerre. Car hire is available.

01 DOMAINE ALPHONSE MELLOT

For 19 generations, as far back as 1513, the Mellot family has been making wines here. This incredible history is apparent as you wander through the 15th-century cellars beneath the streets of Sancerre. Tasting the wines, though, alerts you to the advances of modern winemaking. The latest Alphonse, known as Junior, has made this 50-hectare (123-acre) vineyard biodynamic, limiting each vine to just four to six bunches of grapes at harvest, and sending quality soaring. The cornerstone, La Moussière, is a classic flinty Sancerre, while the barrel-aged Cuvée Edmund has a richness of flavour and subtle aroma that is rare for a Sauvignon. And Generation XIX is a spectacular Pinot Noir, meriting comparison with great Burgundy vintages rather than Sancerre Rouge.

www.mellot.com; tel 02 48 54 07 41; Rue Porte César, Sancerre; Mon–Sat by appointment

02 DOMAINE ANDRE DEZAT & FILS

The Dezat family were pioneers of Sancerre's philosophy to sell directly to the public through estate visits, and the family has a well-deserved reputation for producing outstanding wines. Their signature Sauvignon Blanc is always aged in steel vats, while the excellent Pinot Noir matures in three- to five-year-old barrels, with a Cuvée Spéciale from 50-year-old vines in new oak, perfect if you are patient enough to lay down the wine for a few years. Whatever time of day you pass by for a tasting, it's like dropping in on a party of old friends, with a mix of, say, Dutch tourists who have been buying for 20 years and Parisians visiting to restock their *cave*.

www.dezat-sancerre.com; tel 02 48 79 38 82; Rue des Tonneliers, Chaudoux, Verdigny; Mon–Sat

03 DOMAINE PAUL CHERRIER

Stéphane Cherrier is a young *vigneron* with a lot of respect for the past. A proud portrait of his grandfather in WWI uniform hangs in the tasting room here, and Stéphane recounts how, 'my grandmother used to raise goats to make cheese, which often saved families from poverty when there was a bad harvest or before Sancerre became such a popular wine.' Some of the family's vines grow on the flat in *argilo-calcaire* (limestone and clay) soil, while

01 Onwards to the Châteaux de Chambord

02 Loire canal, Ménétréol-sous-Sancerre

03 Paul-Henry Pellé of Domaine Henri Pellé

those on the slope are marked by the more distinctive *terre blanche* (clay, limestone and oyster shells); it is these two soil types that really typify the wines of Sancerre. While Stéphane's sharp, acidic Sauvignon is aged in steel vats, he is also working to develop the complex Cuvée Philippa in wooden barrels.
Tel 02 48 79 37 28; Chemin Matifat, Chaudoux, Verdigny; Mon–Fri, by appointment Sat–Sun

04 DOMAINE VINCENT GAUDRY

Vincent Gaudry is an artisan winemaker whose ancient cellar may be in the village of Chambre, but the 11-hectare (27-acre) organic-certified vineyard is spread out in parcels across the communes of Sury-en-Vaux, Saint-Satur, Verdigny and Sancerre, 'because I want grapes growing on the key different soils in the region,' explains Vincent, 'flinty silex, *caillotte* (pebbly) and *argilo-calcaire*'. His wines are explosive, especially Constellation du Scorpion, a Sauvignon made from a parcel of 100% silex. 'I want to continue the unique characteristics we have with our wine,' he says, 'to respect our elders who have made Sancerre famous all over the world, and not change for change's sake.'
www.vincent-gaudry.com; tel 02 48 79 49 25; Petite Chambre, Sury-en-Vaux; by appointment

05 DOMAINE MARTIN

Of all the villages surrounding Sancerre, Chavignol is the most picturesque, its medieval houses tightly enclosed by two steep hillsides criss-crossed with vineyards. The slopes are incredibly difficult to work, both in terms of tending the vines and precariously hand-picking during the harvest. Pierrot, as this domaine's young *vigneron* is known, does not need much persuasion to take visitors up to the top of his two prize vineyards: Les Culs de Beaujeu and Les Monts Damnés (the cursed mountains). You'll need a head for heights, as the vineyard drops off like the edge of a cliff. Back in the cellar, the exceptional wines are perfectly paired with the famous Crottin de Chavignol goat's cheese.
Tel 02 48 54 24 57; Le Bourg, Chavignol; Mon–Sat by appointment

06 DOMAINE PASCAL ET NICOLAS REVERDY

Follow a winding road through scenic vineyards to the tiny hamlet of Maimbray, whose 40 inhabitants include 10 winemaking families.

The *famille* Reverdy is one such dynasty: Pascal Reverdy is helped by his two sons, both committed to becoming *vignerons*. Don't be surprised if he begins a dégustation by pouring his fresh, fruity Pinot Noir: 'I feel the Sancerre Blanc is too aromatic to taste first, as afterwards the rosé and red may appear bland.' This used to be a working farm, and the wine cellar resembles a museum, filled with ancient farming tools, while the cosy tasting room looks like the family dining room,

Paul-Henry Pellé welcomes visitors to his state-of-the-art cellar and may whisk them off in his battered old army jeep for a tour

04 Central Sancerre

05 Sancerre townscape

06 Vineyards, Sancerre

with a long wooden table, kitchen and wood-fired stove.

Tel 02 48 79 37 31; Maimbray, Sury-en-Vaux; Mon-Sat

07 DOMAINE HENRI PELLÉ

The sprawling Pellé vineyard spreads across Sancerre into the adjoining appellation of Menetou-Salon. There are many reasons this is a must-visit winery. It provides the perfect opportunity to judge the Sauvignon Blanc and Pinot Noir from Menetou-Salon, so long the poor cousin to Sancerre, but now the rising star. Paul-Henry Pellé welcomes visitors to the state-of-the-art cellar and, at the drop of a hat, will whisk you off in his battered old army jeep for a tour of the surrounding vineyards. And be prepared to taste a lot of wines too, as Paul-Henry vinifies each clos of vines separately and often bottles them as individual cuvées, showing how the soil can completely change a wine. 'Yes, I have a lot of different cuvées,' he says, 'but that is what is exciting about making wines – otherwise I would just get fed up.'

www.domainepelle.com; tel 02 48 64 42 48; Morogues; Mon-Sat

ESSENTIAL INFORMATION

WHERE TO STAY

MOULIN DES VRILLÈRES

Winemaker B&Bs are rare in the Sancerre region, but visitors here get a warm welcome from Christian and Karine Lauverjat, who provide a full tour of their cellar and a tasting. *www.sancerre-online.com; tel 02 48 79 38 28; Sury-en-Vaux*

LA CÔTE DES MONT DAMNÉS

Jean-Marc Bourgeois is the son of one of the most famous Sancerre winemakers, but chose to become a chef before returning home to renovate an old hotel. Today, his guests can relax in designer rooms and dine on refined tasting menus in his gourmet restaurant. *www.montsdamnes.com; tel 02 48 54 01 72; Place de l'Orme, Chavignol*

WHERE TO EAT

RESTAURANT LA TOUR

This restaurant showcases the many talents of chef Baptiste Fournier. Don't miss the *pigeonneau de St Quentin* (pigeon served with grapes and wild mushrooms). *www.latoursancerre.fr; tel 02 48 54 00 81; 31 Nouvelle Place, Sancerre*

AU P'TIT GOÛTER

A brilliant village bistro with wines from more than 50 local Sancerre producers – the ideal accompaniment to both Crottin de Chavignol cheese, made by the owner's son, and another local speciality, a *friture* of tiny deep-fried fish caught in the Loire. *Tel 02 48 54 01 66; Le Bourg, Chavignol*

06

WHAT TO DO

Sancerre overlooks the mighty Loire River, so for a healthy break from wine tasting, go hiking or cycling along its banks, rent a canoe or, at the sand flats of the village of Saint-Satur, sunbathe at the water's edge (though beware of currents).

CELEBRATIONS

Nothing goes better with a crisp, chilled Sancerre than a juicy oyster. On the last weekend of October, the Fête de l'Huîtres, a huge wine and oyster festival, is held in the immense Caves de la Mignonne, an underground quarry dating back to the 14th century (and otherwise usually closed to visitors). *www.caves-de-la-mignonne.com*

01

[France]

PROVENCE

Renowned for its rosé wines, Provence reveals red and white surprises in its vineyards along the glamorous Côte d'Azur and bucolic countryside lanes.

Provence has a strong claim to be France's oldest wine-producing region, with the cultivation of vines dating back to the arrival in Marseilles of the Phoenicians from Greece in 600 BC. Today it is certainly one of the most attractive when it comes to oenotourism, with many estates offering not just tastings but accommodation in grandiose châteaux or wooden-beamed farmhouses, food pairings in casual bistrots or vineyard picnics, and wine festivals stretching through the long hot summer. Add in seductive Provençal landscapes of lavender fields, olive groves and the mythical golden sandy beaches of the Côte d'Azur, and a trip here is irresistible.

Historically known for rosé, now a hugely popular tipple, Provence's modern *vignerons* are now offering a range that extends from light, aperitif-style rosé through to what is known as 'gastronomic rosé' – a far more full-bodied and often oak-aged wine, perfect to accompany herb-encrusted roast lamb, rabbit stewed in olives and tomatoes, grilled prawns or sardines. Winemakers are also aiming for high-quality reds, blending local Carignan, Cinsault, and Mourvèdre grapes with robust Syrah and Grenache.

The vineyards of Provence spread along the Mediterranean from the hills above Toulon towards St-Tropez and Fréjus in one direction, and as far as Marseilles and Cassis in the other. The mountainous *arrière-pays* provides a sharp contrast in landscape and wines, from the wild Massif des Maures to the stark rocky outcrops of Les-Baux-de-Provence. Expect to meet adventurous *vignerons* who are not just moving towards organic and biodynamic cultivation, but also experimenting with the ancient technique of ageing in terracotta amphoras, offering vegan and natural zero-sulphite wines, and using horses to plough vineyards and grazing sheep to replace chemical weedkillers.

GET THERE
Toulon-Hyères airport is 50km (31 miles) from Saint-Cyr-sur-Mer; Marseille-Provence is 65km (40 miles) away. Both have car hire.

01 DOMAINE D'ESTAGNOL

Bandol is the most prestigious appellation in Provence, produced above Toulon and Marseilles, and prices can be steep at such renowned wineries as Château de Pibarnon and Domaine Tempier. The picture is very different at down-to-earth Domaine de l'Estagnol, where sixth-generation *viticultrice* Sandrine Féraud has probably the smallest Bandol vineyard, just 1.5 hectares (3.7 acres). A former rugby player, Sandrine has a tiny cellar where she uses a mix of traditional cement vats and innovative cone-shaped barrels. 'Each vine can see the sea,' says Sandrine, 'as the Mourvèdre grape used in Bandol should traditionally enjoy a sea breeze.' Her 2015 Bandol Rouge is reasonably priced, while the bargain Vin de France bag-in-box is perfect for a picnic.

www.domainedelestagnol.com; tel 06 01 01 35 52; 1426 Route de la Cadière, Saint-Cyr-sur-Mer; daily by appointment

02 DOMAINE COURTADE

Just across from the Bandol appellation lies the beautiful Mediterranean island of Porquerolles, a protected national park with secluded beaches, dense woods and three thriving vineyards. After WWII, the park authorities decided to uproot a small part of the island's forests and replant with vines, originally intended to act as a fire barrier.

A speedy ferry connects Porquerolles' busy port to the mainland, and a 15-minute stroll beneath fragrant eucalyptus trees brings you to the island's largest winery, Domaine Courtade, stretching over 35 hectares (86 acres) along the coastline. The Carmignac family took it over in the late 1980s, and their impressive rosé, white and red Côtes de Provence wines have been officially organic since 1997. The fruity, fresh rosé is irresistible on a hot afternoon. A one-hour visit includes a vineyard tour, explanation of winemaking techniques in the cellar, and a tasting. Pop in to the Villa Carmignac next door to see the family's impressive modern art.

01 Cassis

02 The cliffs of Calanque d'En-Vau

03 Provence vineyards

04 Harvest time at Domaine La Tourraque

'Each vine can see the sea, as the Mourvèdre grape used in Bandol should traditionally enjoy a sea breeze.'

–Sandrine Féraud, Domaine d'Estagnol

www.lacourtade.com; tel 04 94 58 31 44; Chemin Notre Dame, Île de Porquerolles; daily by appointment

03 DOMAINE LA TOURRAQUE

Though Ramatuelle is just down the coast from glitzy St-Tropez, beyond the jet-set beaches, the scenery gets wilder as you enter the protected headlands of Les Trois Caps. Right at the end of the road (more of a deeply rutted track) lies one of the most spectacular vineyards in Provence. Domaine La Tourraque stretches over 38 hectares (94 acres), with two exceptional segments: one running right to a precipitous cliff edge, looking down on crashing waves; the other within touching distance of a sandy beach. While tastings are free, the vineyards are only accessible on an organised one-hour hike costing €15pp. Ask to taste the organic Harmonie Rouge, an intense almost chocolatey blend of Syrah and Mourvèdre.

www.latourraque.fr; tel 04 94 79 25 95; 2444 Chemin de la Bastide Blanche, Ramatuelle; Mon–Sat

04 DOMAINE DES TROIS CHÊNES

Head off the beaten track to discover the small estate of artisan *vigneron* Régis Scarone, hidden away in the Vallée des Borrels. Cultivating vines that are over 50 years old, he has firm ideas about his winemaking. The grapes are picked by hand, with the white and rosé aged in old-fashioned cement vats, and a mix of old and new barrels used for the red. Although he has virtually eliminated chemical treatments in the vineyards, Régis steers clear of the official organic certification, because 'there is just too much bureaucracy'. You'll be hard pressed to find better value and quality wines than his crisp Inspiration Rosé or Character Rouge, a complex blend of Cinsault, Grenache and Mourvèdre grapes, barrel-aged for 14 months.

www.domaine-trois-chenes.com; tel 06 86 86 60 42; 4619 Chemin des Troisièmes Borrels, Hyères; Mon–Sat by appointment

05 DOMAINE CROIX-ROUSSE

Christophe Durdilly's winery is an under-the-radar address at the end of a dusty track, so call ahead for directions. His wines are mostly classified outside the official

05 Château Matheron

06 Chilled Provence rosé

07 Provençal olives

appellation, sold as Vin de Pays. But once you taste the stellar selection of this maverick sommelier-turned-*vigneron*, you'll realise it is well worth the detour. Christophe has been working this small 7-hectare (17-acre) estate – with no expansion plans – since 2005, maintaining 70-year-old bush vines of Carignan and Mourvèdre, which most other winemakers would have dug up and replanted. Wines are organic, with minimum sulphite, so be prepared for an explosive fruitiness, especially in his Croix-Rousse Rouge.
www.domainecroixrousse.com; tel 06 11 86 93 80; 304 Chemin de Merlançon, Puget-Ville; daily by appointment

06 CHÂTEAU MATHERON

Paul Bernard's modern winery sits on the legendary Route Nationale 7, but if you look up the hill, past the symmetrical lines of vines, you can glimpse the family château, where he has opened two B&B rooms. There is a stunning panorama from the terrace, the perfect spot to sip the estate's outstanding and excellent-value wines. Paul is one of Provence's up-and-coming *vignerons* – be sure to try both his signature Tradition Rosé, ideal for a summer sundowner, and the powerful Syrah-dominated blend of Prestige Rouge, a good match for a rich daube Provençale (beef stew).
www.chateau-matheron.com; tel 04 94 73 01 64; 400 Chemin du Domaine de Matheron, Vidauban; Mon–Sat

07 DOMAINE DE LA FOUQUETTE

The Daziano-Aquadro family run this welcoming winery that combines organic wines with delicious food and a B&B chalet. In the shadow of the looming Massif des Maures mountains, Jean-Pierre cultivates 14 hectares (35 acres) of vines, which yield an award-winning rosé, Cuvée Pierres de Moulin. Take a drive up the hillside and you are soon lost in a thick wood of chestnut, pine and cork oak trees, eventually reaching the cosy chalet with guest rooms, where meals can be provided. Prepare for a feast of Provençal dishes, such as slow-cooked chicken with wild herbs, peppers and tomatoes – accompanied, of course, by the family's excellent wines, perfected over three generations of working the vines.
www.domainedelafouquette.com; tel 04 94 73 08 45; Route de Gonfaron, Les Mayons; Mon–Sat

ESSENTIAL INFORMATION

WHERE TO STAY

LES PIERRES SAUVAGES
An idyllic B&B with five rooms and a swimming pool, overlooking vineyards and lavender fields. Knowledgeable owner Gabrielle organises trips to local vineyards by bike or in her trusty Citroën 2CV.
www.lespierressauvages.com; tel 07 60 39 72 57; Chemin Magnau, Besse-sur-Issole

HOTEL LIDO BEACH
For a break from the wine-tasting trail, enjoy some sun and sea at this delightfully kitsch 1960s hotel with its own private beach, right opposite the island of Porquerolles.
www.lido-beach.com; tel 04 94 01 43 80; 5 Avenue Emile Gérard, Hyères

DOMAINE DE VILLEMUS
Escape to this natural hideaway, just north of Aix-en-Provence, surrounded by olive groves, vineyards, lavender fields and almond orchards. Organic products on offer include homemade olive oil.

07

www.villemus.com; tel 06 22 32 44 84; Jouques

WHERE TO EAT

TERRE DE MISTRAL
In the wild Provençal countryside outside Aix-en-Provence, buffeted by the Mistral wind, the *ferme-auberge* (farmhouse inn) of this dynamic winery offers local cuisine, wine and olive oil tastings, and locavore picnics of farm produce to take away.
www.terre-de-mistral.com; tel 04 42 29 14 84; Route de Peynier, Rousset

LA TABLE DE POL
Taste a terrific choice of Provençal wines in the back-room bar with a plate of cheese and charcuterie, or out on the sunny terrace alongside *petits farcis* (stuffed courgette, aubergine and tomatoes) or the Friday special: *aïoli*, thick chunks of cod, vegetables and a wickedly garlicky mayonnaise.
www.facebook.com/LaTabledePol; tel 04 94 47 08 41; 18 Blvd Georges Clémenceau, Lorgues

LE KIKOUIOU
Brilliant beachside joint proving that even in St-Tropez not everything has to be eye-wateringly expensive. Set between a fragrant pine forest and lush vineyards, with the beach two minutes' walk away, this simple wooden *cabane* offers grilled steaks or fish, and tasty pizzas with crisp salads. The wine comes direct from the owner of the adjacent vines.
Tel 04 94 79 83 94; Route de Bonne Terrasse, Ramatuelle

WHAT TO DO

CARRIÈRES DE LUMIÈRES
One of the world's most spectacular *son et lumière* shows is held in this cathedral-like limestone quarry near the equally dramatic village of Les-Baux-de-Provence.
www.carrieres-lumieres.com

LES CALANQUES
Explore the maze of creeks between Marseilles and Cassis, either by a serious coastline hike or on board a pleasure boat.

CELEBRATIONS

Visit St-Tropez on 28 June for International Rosé Day; rosé tastings, food stalls, concerts and pink illuminations at night.
www.rose-day.fr

01

[France]

THE RHÔNE

From a phenomenal region of France, with snowy mountains in its north and broad, hot valleys to the south, come blockbuster red wines that will dazzle your palate.

The wine region of the Rhône Valley stretches from just below Lyon, past Avignon and right down through the south, where the mighty river meets the Mediterranean. Grapes have been grown here for more than 2000 years and there is a tremendous variety of wines to discover as you travel the length of the valley.

The Northern Rhône, from Vienne down to Valence, boasts spectacular scenery, with the river's steep banks covered by terraced vineyards, producing some of France's most famous wines: the intense Syrah of Côte-Rôtie, Cornas and Hermitage; and the elegant Condrieu made from the complex Viognier grape. Below Valence the landscapes become more Provençal, and Syrah is grown alongside Grenache, Mourvèdre and Carignan. These grapes are often blended, which can produce the potent and celebrated Châteauneuf-du-Pape; up-and-coming appellations such as Gigondas, Vacqueyras and Rasteau; or Côtes du Rhône, the classic wine served in every French bistro. Travellers will quickly discover that winemakers in the northern Rhône tend to be more traditional. But once the road heads south of Valence – where plots of vines are much cheaper – a new generation of younger *vignerons* is moving in, eager to experiment, especially with natural wines, which are sweeping the fashionable wine bars of European cities.

Wine tourism has become a well-organised art in the Rhône, with the traveller offered tempting places to stay on many domaines, while restaurants have woken up to the wonderful opportunities of wine pairings – a sharp white Crozes-Hermitage with local cheeses such as a creamy Saint-Félicien and tangy Picodon; the flinty Saint-Péray accompanying salt-baked line-fished sea bass; and a robust Cornas perfect with a lean fillet of wild venison and forest berries.

GET THERE

Lyon is the nearest major airport, 49km (30 miles) from Chavanay. The train from Paris to Lyon takes 1hr 57min. Car hire is available.

01 Syrah vines above Cornas

02 Cave de Tain vineyards

03 En route to the press

04 Tain-l'Hermitage

01 DOMAINE DU MONTEILLET

Stéphane Montez is one of those classic larger-than-life French winemakers. His state-of-the-art cellar, perched high above the Rhône, is a lively rendezvous where local winemakers, wine merchants, chefs and curious tourists bustle in and out all day to taste his splendid wines. Be sure to go to the back of the tasting room, as a glass wall lets you peek into a barrel cellar carved into the rock face. Stéphane is a 10th-generation winemaker, with vines in the two most prestigious Rhône appellations, the red Côte-Rôtie and white Condrieu. He explains how, till the early 1980s, 'Condrieu was just known as "Viognier", as this was our own native grape, for centuries grown only here. Today, the whole world seems to be planting Viognier, from Australia to California and Chile, but it only becomes a great wine here.' *www.montez.fr; tel 04 74 87 24 57; 7 Le Montelier, Chavanay; Mon–Sat by appointment* $

02 CAVE DE TAIN

Founded in 1933, the Cave de Tain has 300 *associés vignerons*, covering 1000 hectares (2471 acres) of vines, with a reputation that sees critics nominate it as France's top winemaking cooperative. Although it produces a staggering five million bottles a year, membership is restricted to winemakers within a radius of roughly 15km (9 miles), and Tain is one of the rare cooperatives to own its own domaine. These precious vines include highly prized parcels that make it the second-largest owner of Hermitage, which is the name not only of the appellation, but also of the hill that looms over its cellar. A visit here includes a tour of the ultra-modern cellars, a €10 million investment that ranges from a barrel room of 2000 *barriques* (wine barrels made of new oak) to the modern technology of cement 'hippos' used for single-parcel vinification. *www.cavedetain.com; tel 04 75 08 20 87; 22 Route de Larnage, Tain-l'Hermitage; May–Aug daily, Sep–Apr Sat–Sun and by appointment* $

03 DOMAINE COURBIS

Driving into medieval Châteaubourg, you can't miss

'Today, the whole world seems to be planting Viognier, from Australia to California and Chile, but it only becomes a great wine here.'

-Stéphane Montez, winemaker

a huge mural advertising the St Joseph and Cornas wines made by the Courbis brothers, Laurent and Dominique. They can trace their family roots here back to the 16th century, though their modern cellar makes use of all the latest technology.

The vineyards of St Joseph stretch for about 50km (31 miles) along the Rhône, and both the red and white vintages produced at Courbis are reasonably priced and need little further ageing. The Syrah is peppery and rich, while the white Marsanne is incredibly mineral, which is no surprise if you drive up to the Les Royes vineyard and see the barren rocky limestone that the vines shoot up from.

www.vins-courbis-rhone.com; tel 04 75 81 81 60; Route de Saint-Romain, Châteaubourg; Mon-Fri, by appointment Sat

04 DOMAINE ALAIN VOGE

Alain Voge is a highly respected figure in the recent history of Cornas wine. Though officially retired, Monsieur Voge lives next door to his cellar, and pops in most days to give his opinion. The estate is certified organic and virtually 100% biodynamic, a rare achievement in this part of the Rhône. It's overseen by local winemaker Lionel Fraisse. 'Just drive up to the walled terraces of the Cornas vineyard,' he suggests, 'and you will see that our work is just as

05 Laurent and Dominique Courbis of Domaine Courbis

06 Canoeing on the Ardèche River

much that of a builder, spending months each year restoring and repairing the walls that keep the vineyard together. These walls date back to Roman times.' The tasting room is decorated with distinctive modern art, and visitors settle in around a big wooden table as Lionel begins opening bottles.
www.alain-voge.com; tel 04 75 40 32 04; 4 Impasse de l'Équerre, Cornas; Mon–Fri, by appointment Sat

05 DOMAINE DU TUNNEL

Sitting in the comfy leather armchairs of his tasting boutique on Saint-Péray's high street, affable *vigneron* Stéphane Robert asks if many winemakers can claim their cellar is housed in a genuine 19th-century train tunnel. He began making wine in his parents' garage, then persuaded the town hall to sell him an abandoned tunnel. Visiting his tunnel today is a spectacular experience, a 150m-long (492ft) cellar carved into the hillside, where he vinifies, stores barrels for ageing, and receives guests for special tastings, by appointment, well worth the €12 fee. Stéphane produces individual, high-calibre wines of little-known white Saint-Péray and intense Cornas vintages, some from vines over 100 years old.
www.domaine-du-tunnel.fr; tel 04 75 80 04 66; 20 Rue de la République, Saint-Péray; tasting room Mon–Sat, by appointment Wed; tunnel by appointment $

06 LE MAZEL

This corner of the southern Rhône is something of a Holy Land for crusaders of the natural wine movement, with a band of New Age *vignerons* making zero-sulphite wines that may sometimes be unstable, slightly oxidised or a little fizzy, but when perfectly made, will surprise even the most expert taster. Siblings Gérald and Jocelyne Oustric inherited an estate whose grapes used to go straight to the village cooperative. Gérald, however, was intent on making and bottling his own unique wines. In his murky cellar, an ancient stone cottage in picturesque Valvignères, be sure to taste the Cuvée Charbonnières, a distinctive interpretation of Chardonnay, aged for one year in steel vats followed by two years in old wooden barrels.
Tel 04 75 52 51 02; Valvignères; Mon–Sat by appointment

07 MAS DE LIBIAN

As the Rhône heads south below Valence and Montélimar, the winemakers are young, unconventional and pushing boundaries. In the dreamy village of Saint-Marcel d'Ardèche, the Mas de Libian is a matriarchial family of *vigneronnes* whose estate of venerable bush vines dates back to 1670. Hélène Thibon, together with her *maman* and sisters, produces certified organic and biodynamic wines, and ploughs the soil with Bambi, a noble workhorse. Their most popular cuvée is Vin de Pétanque, a blend of Grenache and Syrah eminently drinkable on a steamy summer evening.
www.masdelibian.com; tel 04 75 04 66 22; Quartier Libian, Saint-Marcel d'Ardeche; Mon–Fri by appointment

ESSENTIAL INFORMATION

WHERE TO STAY

LA GERINE

Perched high above the Rhône, with a relaxing pool and spectacular views, this comfortable B&B is surrounded by the vineyards of Côte-Rôtie. *www.lagerine.com; tel 04 74 56 03 46; 2 Côte de la Gerine, Ampuis*

HOTEL MICHEL CHABRAN

An old-fashioned but charming inn on the Route Nationale 7, which travels down to the south of France. It's run by a Michelin-starred chef. *www.chabran.com; tel 04 75 84 60 09; 29 Avenue du 45ème Parallèle, Pont de l'Isère*

DOMAINE NOTRE DAME DE COUSIGNAC

Winemaker Raphaël Pommier and his American wife, Rachel, welcome guests to a rustic farmhouse, hosting tastings of their organic wines every evening. *www.domainedecousignac.fr; tel 04 75 54 61 41; Quartier Cousignac, Bourg-Saint-Andéol*

WHERE TO EAT

AUBERGE MONNET

This romantic restaurant on an island in the Rhône serves regional specialities, such as frogs' legs, stuffed pig's trotters, tasty cheeses and charcuterie. Eric, the welcoming owner, has a brilliant selection of wines sold by the glass. *www.auberge-monnet.com; tel 04 75 84 57 80; 3 Place du Petit Puits, La Roche-de-Glun*

LA TOUR CASSÉE

A cosy village bistro that mixes traditional Ardèche favourites (hearty cabbage soup) with exotic recipes such as a tagine of duck confit with dates and quince – and an excellent list of natural wines. *Tel 04 75 52 45 32; Valvignères; open May–Aug*

LA FARIGOULE

Overlooking a vineyard, this old-fashioned auberge is perfect for kicking back over a chilled Côtes du Rhône rosé accompanied by a delicious *caillette*, the local take on meat loaf, or a salad starring warm goat's cheese. *www.auberge-lafarigoule.com; tel 04 75 04 02 60; Bidon*

WHAT TO DO

From Vallon Pont d'Arc, you can head off for the day on a scenic guided canoe trip along the Ardèche River, during which you will weave through spectacular gorges, and glide under the Pont d'Arc bridge.

CELEBRATIONS

Two wonderful festivals are hosted at different ends of the Rhône: Jazz à Vienne (*www.jazzavienne.com*) for two weeks from the end of June, and Avignon's Theatrical Festival (*www.festival-avignon.com*) running through the month of July.

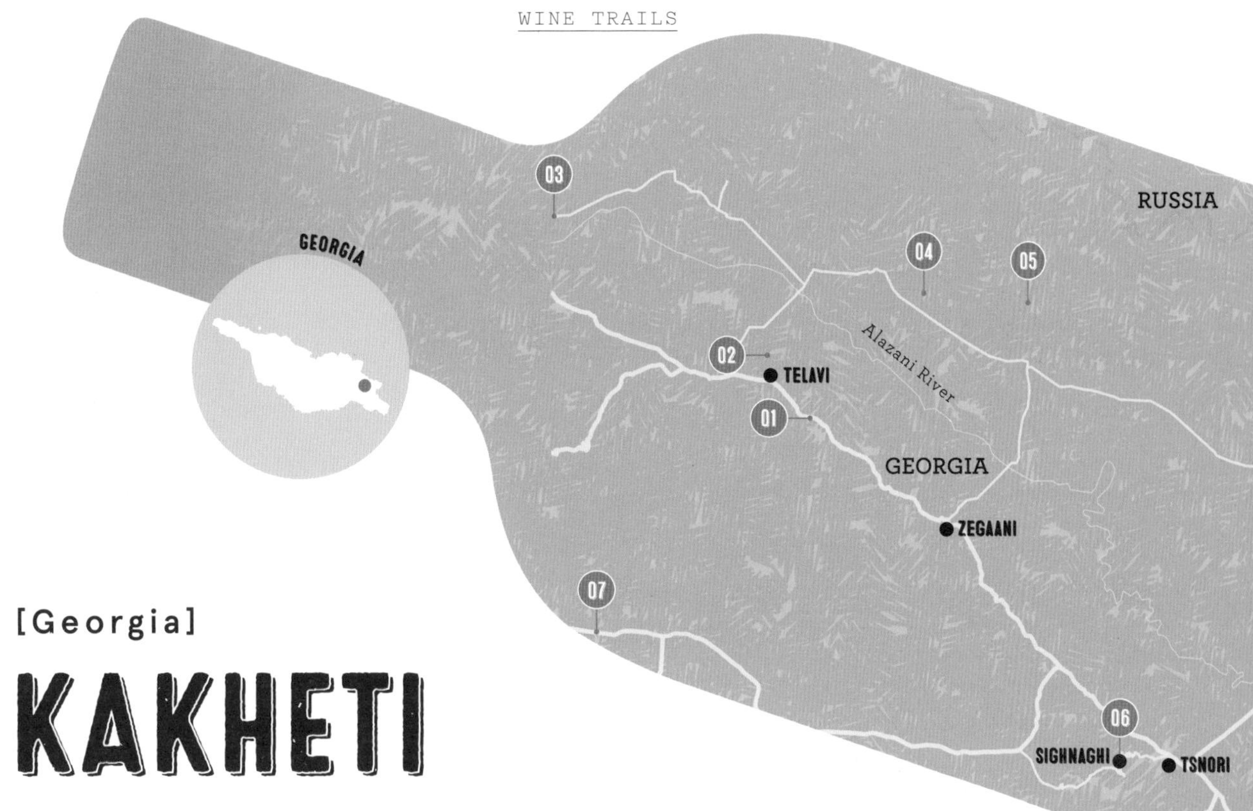

[Georgia]

KAKHETI

In this ancient, mountainous land, a youthful spirit and world-class traditional wines reward grape-loving travellers with a taste for the unknown.

Vineyard-hopping in Georgia is a journey to wine's earliest origins. An 8000-year-old clay wine jar unearthed here in 2017 is the oldest-known relic of winemaking. Put another way, by the time the ancient Greeks were stomping their first grapes, the Georgians had been at it for millennia.

Traditional Georgian winemaking has changed surprisingly little since antiquity: grapes are harvested by hand and foot-pressed in *satsnakheli*, hollowed-out tree trunks. The juice flows into underground *kvevri* clay pots, where it ferments and matures with minimal intervention. The following spring, the wine is clear, aromatic and ready for clinking at *supras*, Georgian feasts known for their elaborate toasts. Kvevri wines are so laborious that they account for less than 10 per cent of the country's wine production, but a new generation of winemakers hopes to turn that ratio on its head.

The most storied Georgian wine region is Kakheti, two hours east of Tbilisi by car, where eighty percent of the country's wine originates. Its best 'whites' are prized for the grippy tannin and amber hue resulting from extended contact between grape juice and skins. But Kakhetian wines are far from homogeneous, thanks to the region's diverse terrain, distinctive grape varieties and innovative winemakers.

The Gombori mountains roughly split Kakheti into 'Inner' and 'Outer' zones. Perched on a Gombori escarpment, Sighnaghi is a popular base, with wineries lining its cobblestoned streets; out in the countryside, there are château-style resorts and countless guesthouses. But Kakheti's most thrilling wines are often found in villagers' backyards, not in sleek tasting rooms. Take time for leisurely meals, impromptu polyphonic singing, chats with winemakers and unsolicited shots of *chacha* (120-proof grape spirit).

GET THERE

Tbilisi International Airport, 100km (62 miles) from Sighnaghi, services the region. Car hire is available.

01 TOGONIDZE'S WINE CELLAR

At first glance, Togonidze's Wine Cellar could be mistaken for an art commune, with its walls adorned with abstract paintings and tables covered in colourful doodles, but down in the cellar, artist-turned-winemaker Gia Togonidze makes some of the region's most rave-worthy kvevri wine. A whiff of his Mtsvane will almost knock you off your chair with its forceful nose of overripe apricots and toasted walnuts, while his Saperavi features earthier notes like mushrooms and coffee. It'd be remiss not to stay for dinner – Togonidze's wife, Lika, is a local celebrity for her modern riffs on Georgian dishes like aubergine *pkhali*, a traditional vegetable-walnut spread that she enriches with caramelised onions.

www.facebook.com/togonidzeswine; tel 591 22 95 94; Shalauri, Telavi; daily by appointment

02 MARANI RUISPIRI

Ruispiri has quickly achieved cult status in the natural wine world for its commitment to biodynamic winemaking, which goes beyond conventional organic farming to incorporate mystical practices like burying manure-filled cow horns in the vineyard, and planting and harvesting according to the cycles of the moon – and the only 'pesticides' used on the vines are herbal teas and lavender. The winery has an enjoyably eco, Burning Man-like vibe with rainbow-painted sheds, abandoned wooden pallets, and a graffitied jalopy parked out front. The wines, however, are elegant and precise, a testament, perhaps, to winemaker Georges Aladashvili's Swiss training. Sample the red Rkatsiteli, a virtually unheard-of varietal redolent of caramel-dipped apples. Farm-to-table meals overlooking the vines can be arranged by advance booking.

www.ruispiris-marani.com; tel 557 51 11 55; 12 Napareuli, Ruispiri, Telavi; daily

03 LAGAZI WINE CELLAR

One of Georgia's most promising young winemakers is Shota

01 Vineyards around Alaverdi Monastery

02 Kakheti countryside

03 Juso's Rkatsiteli and Saperavi

04 Juso's Winery

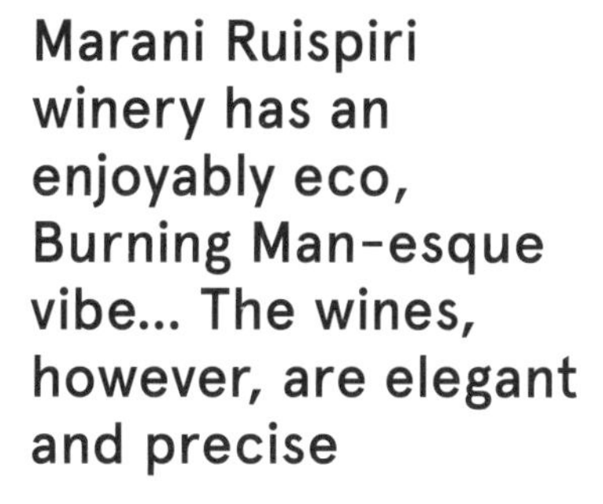

Marani Ruispiri winery has an enjoyably eco, Burning Man-esque vibe... The wines, however, are elegant and precise

Lagazidze, a bearded back-to-the-lander who left a career in tourism to pursue ancestral winemaking. Lucky us – his musky Rkatsiteli (amber) and rose-scented Saperavi (red) are so enchanting you'll want to snap up as many as you can; with his annual production clocking in at 2000 bottles, you just might clear him out. In addition to tastings, Shota and his family can whip up a feast of soup dumplings, cheese bread, stews and salads from their native Tusheti, the remote mountainous corner of Kakheti known for its snowbound villages and prehistoric, pagan-influenced culture. Shota, a proud Tush, will happily regale you with fascinating stories from his homeland between toasts.

www.facebook.com/LagaziWineCellar; tel 551 94 02 17; Zemo Alvani, Akhmetis Raoni; daily by appointment

04 ARTANULI GVINO

This 'estate', best known for its lusty Saperavi, is basically a boho hang-out for agriculturalist epicureans. It's tempting to linger here all day, sipping wine under the trees and geeking out with seasoned winemaker Kakha Berishvili to the soundtrack of the rushing Didkhevi River. Berishvili's daughter, Keti, is now making wine of her own under the label 'Gogo Wine'; her Corazón Partido, a Rkatsiteli mono-varietal fermented on its skins for two months, has a punchy maraschino cherry finish.

www.facebook.com/artanuligvino; tel 599 18 11 01; Artana Village, Telavi; daily, appointment recommended

05 CHUBINI WINE CELLAR

It's worth visiting this buzzy new winery for its Eden-like setting

05 Scenic vineyards at Marani Ruispiri

06 Marani Ruispiri wine tasting

07 The traditional Georgian dish of *khachapuri*

alone, hemmed in by the Greater Caucasus mountains on one side and the Gombori range on the other. You can take in both from a picnic table in the front yard, where husband-and-wife team Tornike and Likuna pour generous tastes of *kvevri*-aged Rkatsiteli and Saperavi. The former gets its floral headiness from the addition of Chinuri, a grape from the Kartli region seldom seen in these parts. Call ahead to book a full-on Georgian *supra* lovingly prepared by the young couple.
www.facebook.com/chubiniwine; tel 599 07 04 28; Shilda, Kvareli; daily by appointment

06 OKRO'S WINES

John Okruashvili heads up this winery, which was instrumental in reviving kvevri winemaking in the early 2000s. Okruashvili's multilayered Rkatsiteli has turned many wine critics' heads over the years, but the current vintages of Kisi (a little-known, gorgeously honeyed grape) and Saperavi (the region's inky standby red) are equally standout. All of Okruashvili's wines are sulphur dioxide- and additive-free and fermented in *kvevri*, and they're available at several price points. Of course, they taste best at the winery's top-floor restaurant, where the balcony looks out over the terracotta rooftops of Sighnaghi towards Azerbaijan.
www.okrogvino.com; tel 599 54 20 14; 7 Chavchavadze Str, Sighnaghi; daily

07 JUSO'S WINERY

If your schedule is too tight for an overnight in Kakheti, you can still get a taste of Georgian wine country at Juso's Winery, situated an hour outside Tbilisi in the Soviet time-warpy town of Sagarejo. Juso's three-man team presides over a fledgling 17-kvevri cellar where award-winning Rkatsiteli and Saperavi ferment underground with no additives or industrial yeasts. Sidekicks to the wine usually include pickled *jonjoli*, bladdernut blossoms, and *guda*, a crumbly ewe's-milk cheese aged in sheepskin. With a few hours' notice, winemaker Lasha Khvedelidze can also throw together a traditional Kakhetian barbecue starring *mtsvadi*, kebabs stacked with hunks of local pork and singed to crackly perfection over grapevine embers.
www.facebook.com/jusowinery; tel 595 55 88 57; 24 Gulisashvili Str, Sagarejo; daily by appointment

ESSENTIAL INFORMATION

WHERE TO STAY

GUESTHOUSING

Hotels abound in Kakheti, but staying with a local family ('guesthousing') can be the best way to go. Georgian comfort food, personal recommendations, and affordable driver services are frequent perks; so is homemade wine. Check out *www.booking.com/guest-houses-kakheti* for a rundown of what's available, or ask at any local winery.

TWINS WINE CELLAR

Eight standard and four upgraded rooms occupy the top floor of Twins Wine Cellar, distinguished by its sky-high *kvevri*, the largest in the world. Wake up to crowing roosters and sweeping vineyard views backed by the Caucasus mountains. *www.cellar.ge; tel 595 226 404; Napareuli*

RADISSON COLLECTION HOTEL TSINANDALI ESTATE

Kakheti finally has a world-class luxury hotel in the Tsinandali Estate, opened in October 2018. Pamper yourself with high-thread-count linens, spacious 'rain' showers and rooftop sunbathing (and cocktailing) by the pool. *www.radissonhotels.com; tel 0350 27 77 00; Tsinandali*

07

WHERE TO EAT

PHEASANT'S TEARS

Chef Gia Rokashvili's creative kitchen hinges on foraged vegetables, fresh herbs and local meats. Wine pairings are a must. In summer, the private outdoor patio, equally suited to intimate diners and large groups, provides spectacular views and, often, spontaneous and joyful live music. *www.pheasantstears.com; tel 0355 23 15 56; Baratashvili Str, Sighnaghi*

KAKHETIAN HOUSE VAKIRELEBI

Zakro and Eka Demetrashvili, owners of this little paradise just outside Sighnaghi, greet all guests with the same phrase: 'Make yourself at home.' After sipping a *piala* (clay bowl) of Zakro's homemade wine, let Eka show you how to make *churchkhela*, a walnut-grape juice confection. Then enjoy a Kakhetian feast. *www.facebook.com/vakirelebi; tel 555 77 78 55; Vakiri, Sighnaghi*

WHAT TO DO

MONASTERY OF ST NINO AT BODBE

Close to Sighnaghi, this is one of Georgia's most sacred churches. It was built in the 9th century AD and houses an active convent and the tomb and relics of St Nino, who brought Christianity to Georgia wielding a cross of grapevines bound with her own hair.

INTER GEORGIA TRAVEL

Wine-lovers seeking hyper-local experiences should link up with Kartlos Chabashvili, Tbilisi-based guide and owner of Inter Georgia Travel. A native Kakhetian, he knows everyone who's anyone here and is a terrific translator to boot. *www.intergeorgia.travel*

CELEBRATIONS

Each May, Georgian winemakers flaunt their best bottles at Tbilisi's New Wine Festival, while the Zero Compromise Natural Wine Fair features only natural wine and draws an eclectic crowd of in-the-know oenophiles.

[Germany]

MITTELRHEIN

This small wine region between Rheingau and Mosel often falls between the cracks, but the wines grown along the spectacular Rhine Gorge, a Unesco World Heritage Site, manage to wow almost as much as the scenery.

Nowhere else in Germany are so many castles perched so precipitously on such steep cliffs as can be found along the 65km (40-mile) stretch of the Rhine river from Bingen to Koblenz. Whether you travel north to south or vice versa, the scenery will blow your mind. In the 19th century this part of the Rhine Valley became known as the *Rheinromantik* (Romantic Rhine), when it was a firm stop on the European Grand Tour, and poets and painters such as JMW Turner took inspiration from the wild and beautiful landscape. Today, boat trippers flock to see the Loreleyfelsen, or Loreley's Rock, where the fabled Rhine maiden Loreley sat combing her golden hair, distracting Rhine skippers until their boats ran aground.

With all of this history and folklore, wine can take a back seat, but it has played an important role in shaping the region. In medieval times, the town of Bacharach was a busy reloading port for barrels of wine, and the wealth and splendour of this period can be seen in the town's well-preserved half-timbered houses, narrow streets and ramparts. Koblenz is another important wine town: it is here that the Mosel river runs into the Rhine at the *Deutsches Eck* (German corner). Numerous wine companies are based here, in prime position to source wine from both valleys.

Winemakers love the Mittelrhein, thanks to the region's clear contrast between wine styles. Around Bacharach, vineyards are confined to narrow, cool, lateral valleys of small tributaries, and the Rieslings are tender and light; vineyards facing the mighty Rhine are warmer and benefit from the river's thermal stability, so the Rieslings are more powerful. Though long and thin, the Mittelrhein is conveniently compact, but still offers plenty of space to taste some fine Rieslings in style.

GET THERE
Both Frankfurt and Cologne Bonn Airport are convenient and around 1.5 hours away by car.

01 WEINGUT TONI JOST - HAHNENHOF

Along the cobbled streets of Bacharach, find the tasting room of the Jost family. Look out for the door handle fashioned like a cockerel, made in honour of the vineyard, the Bacharacher Hahn (*Hahn* meaning cockerel). In a lateral valley of the Rhine on Devonian blue slate, the site turns towards the river, meaning there is coolness and warmth in equal measure. Cecilia Jost, who has made the wines at her parental estate since 2008, manages to express this climate beautifully in her wines. Try her Spätburgunder (Pinot Noir) and don't miss the off-dry styles that have terrific balance.

www.tonijost.de; tel 6743 1216; Oberstr 14, Bacharach; by appointment

02 WEINGUT FRITZ BASTIAN

You'll find the estate's tasting room and small wine shop on the cobbled Bacharach main square. If you see CDs for sale it's because the owner Fritz Bastian was an opera singer before turning to the wines of his family estate. Bastian is the lucky owner of an island in the Rhine: the fruit from the sandy vineyards is vinified separately and the wine is sold as Bastian Riesling Insel, or Island Riesling. Most of the plantings are Riesling, but look out too for the expressive, tender Scheurebe wines.

www.weingut-bastian-bacharach.de; tel 6743 937 8530; Koblenzer Str 1/Corner of Rosenstr, Bacharach; Fri–Sun

03 WEINGUT RATZENBERGER

Joachim Ratzenberger loves his vineyards. This normally reserved man comes alive when he talks about his vines in the Bacharacher Wolfshöhle and the Steeger St Jost winegrowing areas, both of which are in the narrow, cool, lateral valleys typical for the southern part of the Mittelrhein. Of course, Riesling plays the main role and a sizeable proportion of it is turned into Sekt (sparkling wine). The Ratzenbergers were amongst the first to revive the old Mittelrhein

01 Burg Gutenfels in Kaub, beside the Rhine

02 Riverside Boppard

03 Phillips-Mühle wine shop, St Goar

Florian Weingart wanted to do things differently: rather than make more wine, he decreased production to concentrate on making better wine

Sekt tradition when they started making traditional method sparkling wines in the late 1980s. These Riesling Sekts have real personality and age effortlessly. It's easy to overlook the still wines when such stellar bubbles are on offer, but both the Riesling and the subtle, lovely Spätburgunder (Pinot Noir) are great value.
www.weingut-ratzenberger.de; tel 6743 1337; Blücherstr 167, Bacharach; by appointment

04 WEINGUT PHILIPPS-MÜHLE

The mill (*Mühle*) in the name of this estate is not just a quaint reference; it still stands, as it has done since 1265, in the narrow Gründelbach valley.

When brothers Thomas and Martin Philipps started their wine estate, most of the vineyards in the vicinity were abandoned. They started slowly and are still consolidating their holdings, but their wines have a great future in store, with modern labels and clean-cut styles, not to mention the obvious ambition of the two brothers. Their tasting room is tiny but they also run a wine shop and cafe on the river in St Goar, just opposite Loreley's Rock (Loreleyfelsen).
www.philipps-muehle.de; tel 6741 1606; estate: Gründelbach 49, St Goar; cafe & shop: An der Loreley 1A, St Goar; by appointment $

05 WEINGUT MATTHIAS MÜLLER

The Müller family has farmed the slopes of the Bopparder Hamm since 1678, growing cherries as well as vines in the past. Marianne and Matthias Müller have extended the vineyards here from 4 to 17 hectares (10 to 42 acres) and both of their sons, Johannes and Christoph, are joining the estate, invigorating it with their ideas and approaches. They have a large tasting room and an extensive offering of Rieslings, which showcase the exposed, steep nature and warm soils of the Bopparder Hamm perfectly. Even their finest wines are great value.

www.weingut-matthiasmueller.de; tel 2628 8741; Mainzer Str 45, Spay; daily

06 WEINGUT WEINGART

The modern, round winery building at the foot of the Bopparder Hamm vineyard is a little out of the way, but worth seeking out. A circular construction, it's the brainchild of Florian Weingart. His family has made wine here for generations, but Florian wanted to do things differently: rather than make more wine, he decreased production to concentrate on making better wine. He now has 11 hectares (27 acres) across various parcels of the Bopparder Hamm area. He is wedded to the idea of *Prädikate*, the six categories of traditional German fine wine, and crafts these styles with much intuition, teasing out Riesling's inherent interplay of acidity and fruit sweetness.

www.weingut-weingart.de; tel 2628 8735; Peterspay 1, Spay; Mon–Sat

04 Bacharach from above

05 Mittelrhein vineyards

ESSENTIAL INFORMATION

WHERE TO STAY

FETZ – DAS LORELEY HOTEL

This contemporary hotel is in a secluded spot on the Loreley bank of the Rhine. It provides stylish but unostentatious accommodation and the cuisine is acclaimed in the area.
www.fetz-hotel.de; tel 6774 267; Oberstr 19, Dörscheid

BURGHOTEL AUF SCHÖNBURG

Submerge yourself fully in the Romantic Rhine by bedding down for the night in this turreted castle. Enjoy exclusive access to the castle garden and the spectacular views. There is also a little museum.
www.hotel-schoenburg.com; tel 6744 93930; Auf Schönburg, Oberwesel

RHEIN-HOTEL BACHARACH

This half-timbered hotel in the heart of Bacharach is a convenient base. Its contemporary rooms with their exposed beams make the most of the old structure.
www.rhein-hotel-bacharach.de; tel 6743 1243; Langstrasse 50, Bacharach

05

WHERE TO EAT

LANDGASTHOF EISERNER RITTER

Local produce is at the heart of this restaurant; there is even a menu themed around the Mittelrhein cherry. Popular with the locals, it also showcases the best wines of the region.
www.eiserner-ritter.de; tel 6742 93000; Peterskirche 10, Boppard

WEINSTUBE 'ZUM GRÜNEN BAUM'

This typical, old-school *Weinstube* in Bacharach is run by the Bastian family of the Weingut Bastian. The oldest part of the building dates from 1421 and simple fare is the order of the day. Think cold cuts alongside Bastian's local wines.
www.weingut-bastian-bacharach.de; tel 6743 1208; Oberstr 63, Bacharach

FEINKOSTMETZGEREI UWE SCHMIDT

Head to this butcher and delicatessen to stock up for a meaty picnic. Try the *Leberwurst* (liver paté) and the *Rheinischer Saftschinken* (local ham). There is also cheese and wine.
www.feinkost-schmidt.de; tel 261 9730 366; Alte Heerstr 34, Koblenz

WHAT TO DO

What with castles, boat tours and hiking trails galore, as well as the spectacular steepness of the Rhine Gorge, it's no surprise that the Mittelrhein can be something of a tourist trap in high season. Nonetheless, do not miss Bacharach, an incredibly well-preserved town of half-timbered houses and quaint cobbled streets.

CELEBRATIONS

Each summer, a festival of fireworks and illuminations takes place in the towns of the Mittelrhein. Castles are lit up against the night sky, and thundering pyrotechnics explode and echo off the cliffs while visitors watch from specially chartered boats. This series of events, ocurring alongside a programme of concerts and parties, is known as Rhein in Flammen (Rhine in Flames). Booking ahead is essential.
www.rhein-in-flammen.com

[Germany]

MOSEL

Vertigo-inducing vineyards, meandering river bends, unparalleled variations on the theme of Riesling and even Pinot Noir; thrill-seekers will feel right at home in the Mosel.

Few wine regions are as dramatic as the Mosel. Named after the river that takes 237km (147 miles) of looping bends to cover the linear 96km (60-mile) distance from Trier to Koblenz, it is Germany's Riesling canyon. It is a region of superlatives encompassing the steepest vineyards, the narrowest valleys, the most breathtaking Rieslings and a string of diverse villages that, to the initiated, read like a wine list: Leiwen, Piesport, Braunberg, Bernkastel, Graach, Wehlen, Traben-Trarbach and Enkirch. But that is just the Mosel – two of its tributaries, Saar and Ruwer, are also subsumed under this region's name.

The three river valleys are quite distinct, and wine fans can compare tasting notes from the fruit-driven Rieslings of the Mosel with the more austere but equally thrilling wines from Ruwer and Saar. The landscapes differ, too: the Saar is wilder and more spacious; the Ruwer tiny and impressive; and the Mosel valley is the most storied. Driving along the river is a must: the road runs right along the bank past countless world-famous vineyards, their names proclaimed in white lettering from afar. Locals have a hard time of it, stuck behind the awestruck tourists creeping slowly along the winding roads, craning their necks to marvel at the steepness of the terrain. But this is an indispensable part of the trip, as the unfolding vistas explain why the spectrum of wine styles is so wide: steepness, altitude, aspect, relative distance from the river all have a bearing on the finished wine, which comes in endless permutations here. Add viticultural differences like planting density and harvest points to this, as well as different winemaking decisions, and the mind starts to boggle. Tasting your way through the resulting abundance of styles is one of the great pleasures of a visit.

GET THERE
Frankfurt-Hahn is the closest airport. Otherwise Luxembourg airport is closer than any German airports. Car hire is available.

01 WEINGUT HEYMANN-LÖWENSTEIN

This tasting room allows for a real flavour experience of the Lower Mosel, the last stretch of the river before it runs into the Rhine at Koblenz. Owners Cornelia and Reinhard Löwenstein were pioneers of authentic Mosel styles when they started out in the 1980s. Today, all the Rieslings that they produce are fermented with wild yeast in old oak *Fuder* barrels. The differences in slate between their chief sites, the Winninger Uhlen and Röttgen, are beautifully clear. Their new winery, a distinctive cube-shaped edifice, is emblazoned with fetching metal lettering that quotes the German translation of Pablo Neruda's poem 'Ode to Wine'.

www.hl.wine; tel 2606 1919, Bahnhofstr 10, Winningen; Fri–Sat & by appointment $

02 WEINGUT MELSHEIMER

This estate has been in the same family for five generations, run organically since 1995 and received biodynamic certification in 2013. But that alone would not make it outstanding. What makes it special is winemaker Torsten Melsheimer's unconventional approach: try the Vade Retro Riesling made without added sulphur, the bone-dry Sekt and the vivid, delicious Pet Nat. This is the Mosel, but not as you know it. Phone ahead for a tasting.

www.melsheimer-riesling.de; tel 6542 2422; Dorfstr 21, Reil; by appointment $

03 IMMICH-BATTERIEBERG

The wine labels at Immich-Batterieberg show little angels with a cannon: a reference to the creation of the vineyards in the 19th century by blasting away the rocks with dynamite. Farmed organically and handcrafted by Gernot Kollmann, try the impressively pure Rieslings of sinuous elegance. If you are lucky, you will find the companionable Gernot himself in the tasting room. If so, you are in for a wonderful chat.

www.batterieberg.com; tel 6541 8159 07; Im Alten Tal 2, Enkirch; by appointment

04 WEINGUT SELBACH-OSTER

On most days, you can drop into this brand new *Vinothek* (wine store) without an appointment.

01 Markus Molitor vineyards

02 Restaurant dining at Weingut Peter Lauer

03 Peter Lauer's riverside vineyards

04 Florian Lauer picking grapes by hand

In the same family for five generations, Torsten Melsheimer's estate is special due to his unconventional approach. This is the Mosel, but not as you know it

Do not let the contemporary architecture deceive you, this estate is famed for its nuanced interpretations of classic Kabinett, that most light-footed of Riesling styles. There is a charge for tasting, but it's worth it for the expert explanations you'll receive from your host. Try comparing the separately vinified parcels of old and age-old vines from the same vineyard – an educational treat! *www.selbach-oster.de; 6532 2081; Uferallee 23, Zeltingen; Mon–Sat, Sun by appointment* $

05 MARKUS MOLITOR

Taste world-class Riesling and fine Mosel-grown Pinot Noir in the beautiful *Vinothek* (wine store) of this 19th-century winery. The setting is beautiful, and the wines exquisite, still made by Markus Molitor himself who is now one of the largest yet most exacting producers in the Mosel. *www.markusmolitor.com; tel 6532 95400-0; Haus Klosterberg, Bernkastel-Wehlen; Mon–Fri, weekends by appointment*

06 WEINGUT NIK WEIS SANKT URBANS-HOF

One of the Mosel's top estates, where you can taste Rieslings from both the Mosel and Saar valleys, comparing the wines side by side. The wines are brilliantly executed, textbook examples for the region, and cover the full spectrum from dry to sweet. Thanks to this estate, which also runs a wine nursery and

05 Nik Weis of Weingut Nik Weis Sankt Urbans-Hof

06 Exploring the Mosel by bike

07 Leiwen

08 Cochem Castle, overlooking the Mosel

propagates cuttings from ancient Riesling vines across the region, a lot of the Mosel's original genetic diversity has been preserved.
www.nikweis.com; tel 6507 9377-0; Urbanusstr 16, Leiwen; Mon–Fri $

07 VAN VOLXEM

This estate, revived in 1999 by Roman Niewodniczanski, has only one aim: to make historically faithful, dry Saar Riesling as it was made in its heyday during the Belle Époque. This has been achieved with thrilling results. Everything, from affordable entry-level varieties to age-worthy, single-site wines comes highly recommended. Its brand-new winery is bedded into the Saar landscape.
www.vanvolxem.com; tel 6501 9477 800; Zum Schlossberg 347, Wiltingen; Tue–Sun

08 WEINGUT PETER LAUER

There are two good reasons to visit this winery: Sekt lovers will find long-aged, late-disgorged Riesling Sekts of exquisite quality, while lovers of traditional Prädikat styles (Kabinett, Spätlese, Auslese), will find exacting, show-stopping versions here. Florian Lauer, the estate's fifth-generation winemaker, has a clear philosophy and his wines enjoy (deserved) cult status in New York wine bars.
www.lauer-ayl.de; tel 658 -3031; Trierer Str 49, Ayl; Mon–Fri

ESSENTIAL INFORMATION

WHERE TO STAY

ROMANTIK JUGENDSTILHOTEL BELLEVUE

A real relic from a bygone age, this hotel offers 35 rooms lovingly restored in the Jugendstil (art nouveau) style. Period detail is combined with modern comfort. *www.beelevue-hotel.de; tel 6541 7030; An der Mosel 11, Traben-Trarbach*

SCHLOSS LIESER

After years or painstaking restoration, this sumptuous Mosel palace opened its doors to visitors in August 2019. Find true splendour in a fabulous setting. *www.marriott.com; tel 6531 986 990; Moselstr 33, Lieser*

GÄSTEHAUS CANTZHEIM

On a quiet side arm of the Saar, at the foot of the Kanzemer Altenberg vineyard, this contemporary guesthouse in a baroque building has been finished to the highest standards. The hosts also make wine and hold various events.

www.gaestehaus-cantzheim.de; tel 6501 607 66 35; Weinstr 4, Kanzem an der Saar

WHERE TO EAT

ZELTINGER HOF

Seasonal, regional food and a great wine menu are at the heart of this operation. If you go in spring, you can see the landlord peeling basketfuls of white asparagus in a sunny spot on the street. There are rooms, too. *www.zeltinger-hof.de; tel 65 32 93 820; Kurfürstenstr 76, Zeltingen-Rachtig*

SEKTSTUUF ST LAURENTIUS

Affiliated to acclaimed Sekt producer St Laurentius, this relaxed wine bar does everything from charcuterie or cheese platters to proper fine dining. Rooms are also available. *www.sektstuuf.de; tel 6507 939055; Euchariusstr 15, Leiwen*

RÜSSEL'S LANDHAUS HASENPFEFFER

Situated in the hills above the Mosel, this establishment offers a choice between fine dining in the main house (Landhaus) or rustic, local specialities in the Hasenpfeffer restaurant. *www.ruessels-landhaus.de; tel 6509 9140-0; Büdlicherbrück 1, Naurath/Wald*

WHAT TO DO

Strike out on foot or take to the water to get a real feel for the local landscape. The Moselsteig is a long-distance hiking trail that runs the entire length of the river in 24 stages. The tourist office also offers guides to various well-posted hiking and cycling tracks. Hiring canoes or signing up for a guided canoe tour are also fun ways to see the sights from the water. *www.moselsteig.de; www.mosel-kanutours.de*

CELEBRATIONS

Don't miss two well-organised wine events: Mythos Mosel in mid- to late May and Saar Riesling Summer in late August. They mobilise the entire region over a weekend of tastings and parties. Tickets include free travel on shuttle buses, which serve the tasting stops at various estates. Taste top wines, find new favourites and make friends with an international, Riesling-loving crowd. *www.mythos-mosel.de; www.saar-riesling-sommer.de*

STADTRUNDFAHRT
Elbe Biergarten
THEATERKAHN
KAHNALETTO
ITALIENISCHES RESTAURANT
Das Theater auf der Elbe
01

[Germany]

SACHSEN

Centred around the architectural marvel and cultural treasure that is Dresden, a revived and burgeoning wine industry is flourishing in Germany's most easterly growing region.

The vineyards of Sachsen are situated on a pretty stretch of the Elbe river, making up for their northeasterly location by catching every last ray of sunshine on mainly west-facing slopes. The region's revival since the German reunification in 1989 is nothing short of remarkable. The rundown remnants of the former German Democratic Republic's collectivised, state-run viticulture have been transformed and returned to the fun-loving spirit of the Saxons.

Sachsen viticulture has a chequered history. In the 12th century, the Medieval Warm Period (c 950–c 1250) saw German winemaking spread north and east, but as the climate cooled over subsequent centuries, the struggling vineyards never attained the economic importance evident in other German regions. Originally terraced in the early 17th century, today's vineyards were kept alive despite frosty winters and world wars. With subsoils of granite and hard, igneous syenite, mostly covered with loess and loam, they are choice spots for vines – and for vineyard walks.

At a latitude of 51° north, you might think that ripening is a problem – but the sunshine is more than sufficient, and the wines rounder and smoother than you might expect so far north. White grape varieties dominate, but Spätburgunder (Pinot Noir) also does well. The most planted variety in Sachsen is Müller-Thurgau, followed by Riesling, Weissburgunder and Grauburgunder. But Sachsen also includes a little speciality: 28 hectares (69 acres) of Goldriesling – a variety producing rounded, easy-drinking wines.

GET THERE
Dresden has an international airport, but services to Berlin are more frequent. The direct train from Berlin to Dresden takes two hours.

01 SCHLOSS WACKERBARTH

This lovely destination estate in the hills of Radebeul is Sachsen's state-owned winery. Once a pleasure palace and dating from 1728, it is now thoroughly modernised, with onsite shop, restaurant and a variety of tours available. Due to its gorgeous setting in the vineyards, it also serves as a popular wedding venue. You might think that being so geared to visitors speaks against the wines – but they are increasingly good quality. Sekt is a particular focus, and this is also the place to taste Goldriesling. A 1956-planted field blend – where various grape varieties are co-planted and co-harvested – is vinified each year. Keep an eye on the website's events page listing fairs and evenings of music and theatre; wine always plays a starring role of course.

www.schloss-wackerbarth.de; tel 351 89550; Wackerbarthstr 1, Radebeul; Tue–Sun (weekends only in winter)

02 WEINGUT KARL-FRIEDRICH AUST

Perched amid the vines of the Goldener Wagen vineyard in Radebeul, Karl-Friedrich Aust's wine estate was one of the first to breathe new life into this lovely corner of the world. Aust loves clambering about the stone-walled terraces of the vineyard and his wines show lovely freshness. There is a tiny restaurant, and a small shop, open at the weekends, where you can pre-order picnic baskets to take up into the sunny vineyard terraces. Aust grows mostly white wines and a smattering of evocative Pinot Noir.

www.weingut-aust.de; tel 351 893 90100; Weinbergstr 10; Radebeul; Wed–Sun

01 Dresden cathedral

02 Schloss Wackerbarth

03 Meissen, flanked by vineyards

04 Frauenkirche, Dresden

05 Radebeul

03 WEINGUT HOFLÖßNITZ

Give yourself some time for this visit: this is Sachsen's oldest wine estate where viticulture has been documented since 1401. Set in a lovely old manor house in the vine-covered hills, the wine estate is owned and run by the town of Radebeul who keep a small wine museum here with a roster of events and concerts. The estate has farmed its vines organically since 1992. This explains why fungal-resistant varieties like Johanniter, Regent and Solaris grow alongside Riesling and Weissburgunder. There is a wine terrace over the summer months offering wines by the glass and simple snacks. The wine shop is open year-round.

www.hofloessnitz.de; tel 351 839 8333; Knohllweg 37, Radebeul; daily

04 WEINGUT SCHLOSS PROSCHWITZ

After the fall of the Berlin Wall in 1989, Georg Prinz zur Lippe, then a management consultant in Munich, went east to reclaim his ancestral estate. Bit by bit he bought back what had been confiscated in 1945 and together with his wife built up Sachsen's largest private wine estate. There is a focus on Pinot varieties and Sekt; management is by South African winemaker Jacques Du Preez. His Spätburgunders (Pinot Noirs) are elegant and worth a trip. There is also a small distillery, a guest house and an event space.

www.weingut-proschwitz.de; tel 3521 76760; Dorfanger 19, Zadel über Meissen; daily

05 WEINGUT KLAUS ZIMMERLING

No other wine estate in all of Germany is as enchanting as this one. Sculptures hewn by Malgorzata Chodakowska (her husband is winemaker Klaus Zimmerling) adorn the building and gardens in striking fashion, their figures lending the whole scene a sensuous atmosphere. The entire setting is dreamlike: located at the foot of the Rysselkuppe mountain and part of the former royal vineyard, the winery and house morph with the landscape. Zimmerling's wines are rounded and concentrated. So tiny are the yields from the 4 hectares (9 acres) of vines here that the Riesling, Grauburgunder, Weissburgunder, Roter Traminer and Gewürztraminer are sold in 500ml bottles. Zimmerling also grows a little Spätburgunder (Pinot Noir), exclusively for sparkling wine. Phone ahead for tastings and plan some more time to wander in the hills.

www.weingut-zimmerling.de; tel 174 210 6812; Bergweg 27, Dresden; by appointment

04

ESSENTIAL INFORMATION

WHERE TO STAY

HOTEL VILLA SORGENFREI

Sorgenfrei translates as 'free from worry' and this small gem of a hotel set amid the villas of genteel Radebeul, close to the vineyards, is a relaxing place indeed. A glass of Sekt on arrival is part of the joy, as is dining in the garden. There are just 14 individually decorated rooms. *www.hotel-villa-sorgenfrei.de; tel 351 795 6660, Augustusweg 48; Radebeul*

SCHLOSSHOTEL PILLNITZ

The Schlosshotel Pillnitz offers the chance to stay in the summer residence of none other than Augustus the Strong. This family-run hotel is set in quiet parkland and close to the vineyards of Pillnitz. The rooms have period charm while the hotel offers a restaurant, a cafe and a beer garden. *www.schlosshotel-pillnitz.de; tel 351 26140; August-Böckstiegel Str 10, Dresden*

HOTEL TASCHENBERGPALAIS KEMPINSKI

Enjoy five-star luxury in the centre of town in the historic setting of the Taschenbergpalais. Originally built by Augustus the Strong for his mistress, the Countess of Cosel, it was destroyed during Dresden's devastating bombing in 1945, but is now a contemporary hotel in a prime location. *www.kempinski.com/en/dresden/hotel-taschenbergpalais; tel 351 49120; Taschenberg 3, Dresden*

WHERE TO EAT

PFUNDS MOLKEREI

Stock up on cheese and other picnic essentials at this ornately tiled, historic dairy. If you prefer a sit-down treat, go upstairs for ice cream or cakes. *www.pfunds.de; tel 351 810 5948; Bautzner Str 79, Dresden*

WEIN.KULTUR.BAR

An exquisite showcase for wines from Saxony and across Germany, all personally selected by owner Silvio Nitzsche. Fabulous cheeses round out the stellar experience. Guests usually leave swooning. *Tel 351 3157917; Wittenberger Str 86, Dresden*

ADAMS GASTHOF

In the Sachsen countryside, this half-timbered inn dating from 1675 is close to the charming Mortizburg castle. Down-to-earth but well-executed Saxon specialities are served in a lovely farm-style setting. *www.adamsgasthof.de; tel 35207 99775; Markt 9, Moritzburg*

WHAT TO DO

Dresden is a cultural treasure trove. Visit the restored Frauenkirche and marvel at Augustus the Strong's extraordinary Zwinger palace. Book ahead for the Semper Opera (*www.semperoper.de*) or simply stroll around the picturesque Old Town.

CELEBRATIONS

Every year on the first weekend of September the wine festival in Weinböhla draws crowds. Grilled sausages are washed down with local wine and the entire village is festooned with vine leaves and garlands. Be prepared for oompah bands and other live music on Friday and Saturday nights. *www.weinboehla.de*

DALAMÁRA
DALAMÁRA
paliokalias
| 2011 |
NAOUSSA
PROTECTED DESIGNATION OF ORIGIN
14%VOL
PRODUCED AND BOTTLED BY DALAMARA WINERY IN NAOUSSA
WINE OF GREECE
750 ML e

01

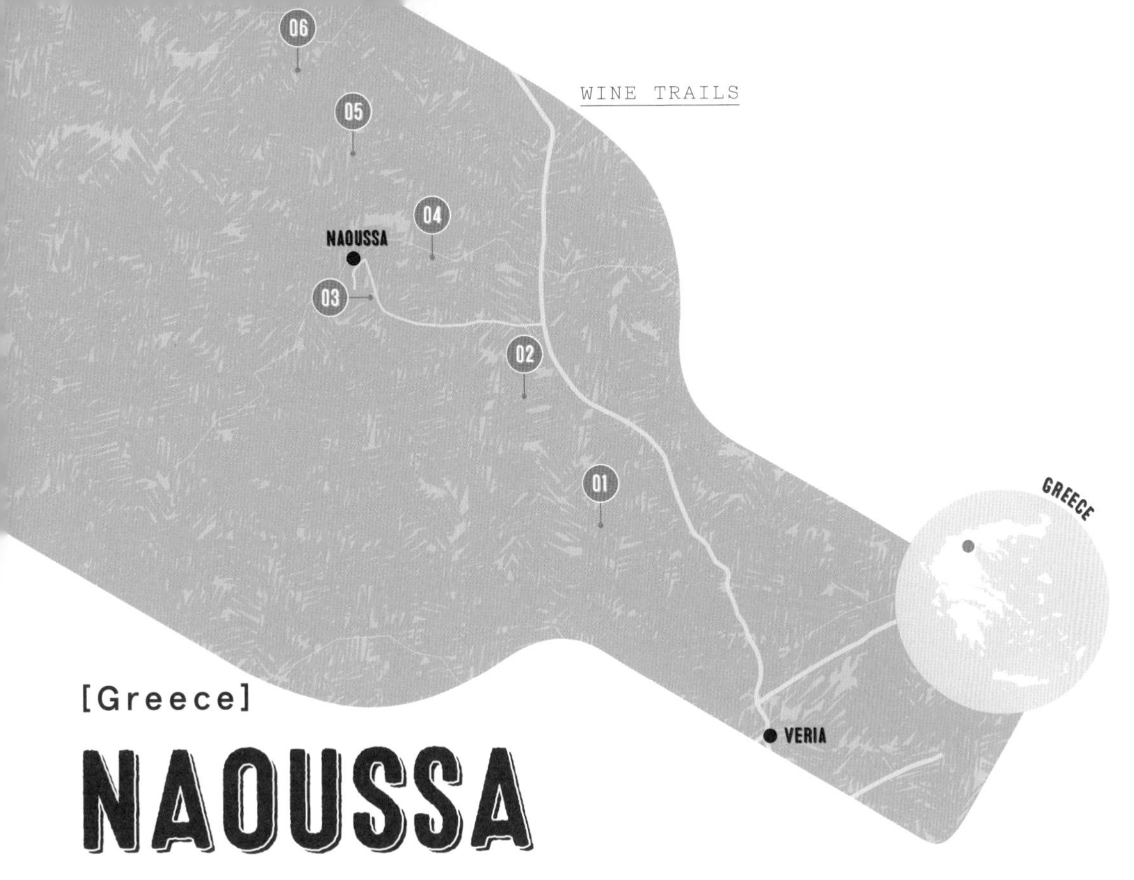

[Greece]

NAOUSSA

Verdant and wild, Naoussa's cool hills attract as many outdoor enthusiasts as they do wine-lovers, who come for a taste of Greece's best red wines.

Drive only an hour west of Thessaloniki and you'll find yourself cosseted in the forested hills of the Vermio Mountains. The streams and waterfalls that course down the deep-green slopes once made the region a powerhouse in the silk trade, with Naoussa at its centre; today, they feed the vines responsible for the country's most exalted red wines.

Those wines are made from just one grape, Xinomavro (pronounced kse-*no*-mav-ro). The name means 'sour-black', though it's not actually sour nor particularly deeply coloured. In fact, its wines are most typically aromatic and perfumed, with plummy fruit filigreed with notes of thyme, saffron, liquorice and sun-dried tomato. Xinomavro is often compared to Nebbiolo for its powerful tannins, acidity and ability to age for decades; the variety also earns comparisons to Pinot Noir for its delicate, detailed flavours.

It also shares a reluctance to travel, a magic synergy with place: as Nebbiolo is to Piedmont, and Pinot Noir is to Burgundy, so Xinomavro is to Naoussa. In fact, in 1971 Naoussa was the first modern winegrowing area in Greece to be recognised for top-quality wine, thus setting a standard for subsequent regulations across the country. If you want to taste great Xinomavro, it's worth the trip to taste it in situ.

The best way to approach Naoussa is from Thessaloniki. Driving west across the plain of Imathia, the Vermio Mountains form a high, green crescent against the sky. Follow narrow, twisty mountain roads up to Naoussa's city park and gaze out over the plains to the east and south. You can see why Naoussa became a crossroads of culture and identity, and take in the wild beauty of a lush winegrowing landscape that's one of Greece's national treasures.

GET THERE

Thessaloniki airport is 93km (58 miles) from Naoussa. Trains run five times a day and take just over an hour; double that for the bus trip.

01 THYMIOPOULOS VINEYARDS

Apostolos Thymiopoulos is one of the most talked about winemakers in the Naoussa region. In 2004, he began to help his father bottle wines from vineyards his family had cultivated for generations. Now at the helm of the estate, he works in the vineyards and winery with as little intervention as possible. His wines speak of the vineyards' purity as well as the warmth and rich soils of Trifolos, in the lower altitudes southeast of town. Bottlings range from the fresh, fruit-forward Young Vines Xinomavro to the full-throttle Uranos, intended for long ageing. He also makes one of the best pink wines in Greece, the earthy, savoury Rosé de Xinomavro.

www.thymiopoulosvineyards.gr; tel 23310 93604; Trilofos, Imathia; by appointment

02 BOUTARI WINERY

For decades, Naoussa wine was synonymous with Boutari, the first winery to bottle a Xinomavro commercially, in 1879. By buying grapes from farmers all over the countryside, the family was instrumental in keeping the grape-growing tradition alive in a region that's had its share of political unrest. The family-run company has grown to include wineries all over Greece, but its heart remains in Naoussa. The winery itself is a landmark: it has one of the largest cellars for ageing wines, with a collection of bottles going back decades. Visits include a multimedia display that tells the story of the region and its wine, a tour of the facilities and tastings of the wines paired with local food. Boutari's Naoussa Grande Reserve remains its flagship wine; also look for the 1879, which is harvested from a single outstanding vineyard in Trifolos.

www.boutari.gr; tel 23320 41666; Stenimachos, Naoussa; Mon–Fri & by appointment

03 DALAMÁRA WINERY

Located on the eastern foothills of Mt Vermio, just outside the town of Naoussa, this historic, organic estate is currently run by Kostis Dalamára, the sixth

01 Tasting at Dalamára winery

02 Greek salad

03 Dalamára vineyards

04 Winemaker Kostis Dalamára

05 Keeping vine leaves for future use at Argatia

06 Thymiopoulos vineyards

generation of the family to farm vines here. Although Kostis studied in Burgundy and has made wine in California, France and Spain, he returned home on the strength of his conviction that Xinomavro is an outstanding grape. He farms his vines as his father did, without chemicals or fertilisers; the oldest date back 100 years. Take time to walk through the vineyards, listening to the chickens cluck in the family's kitchen garden; afterwards, take a quick tour of the winery before adjourning to the cosy, stone-walled space that serves as its tasting room. The star wine here is the oak-aged Paliokalias Naoussa Xinomavro, which balances its rich body with delicate spice scents. Beyond this, of special interest are the 'cellar offerings', or 'confidential cuvées': experimental bottlings of blends, clones and single vineyards offered only at the winery.

www.dalmara.gr; tel 23320 28321; Epar. Od. Naoussa–Kato Vermiou 31, Naoussa; by appointment

04 KARYDAS

Karydas is 'boutique' on all counts. Only one Xinomavro wine is made, the property is tiny, the production is minuscule, and it always sells out. Upon visiting, however, you realise that there's nothing boutique about the estate – in fact, it doesn't feel like an estate at all, but instead, the house of a loving relative who just happens to make extraordinary wine. Petros Karydas currently runs the winery, tending by hand the 2.5 hectares (6 acres) of vines surrounding the house; his vinification methods are just as artisanal, the wines fermented and aged in a small array of cement tanks and old oak barrels in a space under his house. The wine itself is pure Naoussa, herbal and earthy; delicious in its youth, it also improves with age.

www.domainekarydas.com; tel 23320 28638; Ano Gastra, Naoussa; by appointment

05 KIR-YIANNI

Kir-Yianni means 'Sir John'; in Greek, the greeting connotes a special warmth and cordiality. The 'John' here is Yiannis (John)

Boutari, of the famous Boutari winemaking family, who split off from the family business in 1996, taking some vineyards with which to build his own brand. Based in Yianakohori, at one of the highest altitudes in Naoussa, the estate has since become one of the region's most famous, recognisable by the picturesque 200-year-old stone lookout tower in the midst of the vines. In part, the estate's fame has to do with Yiannis himself, who's deeply involved in politics and conservation (he runs a wildlife reserve in nearby Nymphaio, and has served several terms as the mayor of Thessaloniki). But the winery is also an extension of his sense of responsibility to the region: here, working with his sons, Stelios and Mihalis, he's created one of Naoussa's most forward-looking estates, the vineyards dedicated to researching the area's full potential with wine. Kir-Yianni offers a dizzying number of bottlings, including Xinomavro in sparkling, rosé and red versions, as well as in blends with Merlot and Syrah. The tasting-room experience here can be lengthy, yet is terrifically enjoyable and intimate. And don't miss the sublime views: at 300m (1000ft) in altitude, you can see clear across the treetops to the town of Naoussa.

www.kiryianni.gr; tel 23320 51100; Yianakohori, Naoussa; by appointment

06 ARGATIA

Few people know the vineyards of Naoussa as intimately as Haroula Spinthiropoulou and her husband, Panagiotis Georgiadis. Haroula is one of the most respected agronomists in Greece, a sought-after consultant who has worked with many of the region's top producers; Panagiotis was the director of The Wine Roads of Northern Greece, an organisation dedicated to promoting wine tourism in Naoussa. Together with their son, Christofer, they run this small winery out of the bottom of their house in Rodochori, a high, forested area on the northwest edge of the appellation. A visit with them is like getting a masterclass in Naoussa's wines, only with warm, open friends as teachers rather than stuffy professionals. Xinomavro predominates, which Haroula makes in a delicate, layered style. She also grows a range of lesser-known varieties, such as Negoska and Mavrodaphne, which she works into fascinating blends.

www.argatia.gr; tel 23320 51080; Rodochori, Imathia; Mon–Fri & by appointment

ESSENTIAL INFORMATION

WHERE TO STAY

PALEA POLI

Located in the heart of Naoussa's 'Old City', this stone mansion houses eight rooms and one luxury suite. They are all surprisingly affordable (about €90 per night) given the elegant decor and exceptionally good breakfast. The hotel also hosts a wine bar with an extensive array of local bottles. *www.paleapoli.gr; tel 23320 52520; Vassileos Konstantinou 32, Naoussa*

SFENDAMOS WOOD VILLAGE

Sfendamos rents out six rustic-chic chalets in the Vermio Mountains fitted out with fireplaces and free wi-fi; the main chalet functions as a gathering place where guests play board games and dine on traditional dishes. There's plenty of Xinomavro on the wine list, too. *www.sfendamos.gr; tel 23320 44844; Pigadia, Naoussa*

WHERE TO EAT

OINOMAGEIREMATA

The name says it all, translating roughly as 'wine and cooking place'. Just off the town's main square, this cosy space excels at the region's traditional cuisine, including fish that owner Dimitris Tavoularis catches himself. His wine list is one of the town's best, reasonably priced and deep in vintages. *Tel 23320 23576; 1 Dragoumi Stefanou, Naoussa*

06

12 GRADA

A short drive south of Naoussa, 12 Grada stands as one of the region's finest watering holes. The food ranges from simple to fancy, but it's always great, drawing in a steady stream of locals, including winemakers. *www.12grada.gr; tel 23311 00112; Sofou 11, Veroia*

WHAT TO DO

POLYCENTRIC MUSEUM OF AIGAI

It's worth the detour to Vergina to check out the archaeological ruins of Aigai, the first capital of the Kingdom of Macedonia. Now a Unesco World Heritage Site, the ruins, including a lavish palace and extensive necropolis of ancient royals, attest to the prestige and power that the area once enjoyed. *www.aigai.gr*

RIVER ARAPITSA

A short drive from Naoussa, take a walk along the storied Arapitsa, the ancient natural border between the plains of Imathia and the Vermio Mountains. In summer, green forests and waterfalls provide a backdrop for meditation on historical forces that have shaped this region.

3-5 PIGADIA

Naoussa offers some of the best skiing in Greece. The slopes of 3-5 Pigadia range from beginner level to steep black diamonds and off-piste areas. During the summer, you can explore the area on mountain bike. *www.facebook.com/35pigadiaski*

CELEBRATIONS

Carnival is Naoussa's largest celebration: 12 days of springtime festivities capped off with a costume parade, and abundant chances to enjoy Xinomavro. The city also stages 'Wine and Culture' events during the September to November harvest.

01

GREECE

[Greece]

SANTORINI

With its high cliffs capped with whitewashed buildings and deep blue sea, this volcanic isle claims one of Greece's most dramatic landscapes – with wines to match.

Greece is home to many islands worthy of a visit, but none come close to Santorini for sheer drama. A half-moon of land in the middle of the Aegean, it's the remnants of a volcano that blew out its side in Minoan times. The force of the explosion, and the ensuing pyroclastic flows, created a tsunami that knocked out civilisation on the northern shores of Crete, some 70km (43 miles) south. It also created a caldera nearly 400m (1310ft) deep. Filled with sky-blue water and framed by 300m-high (984ft) cliffs, it's a stunning landscape, especially when it's drenched in sunlight – which it is more than 300 days a year.

As gorgeous as it may be, this is a challenging environment for plants, save for Assyrtiko, a grape variety that's the foundation of the island's namesake wine. The vine has thrived here for hundreds of years, developing an uncanny ability to ripen under the island's hot sun without losing its acidity. The resulting white wines nearly quiver with energy, their flavours bright and citrusy, with a mouthwatering salinity. And they are easy to recognise, labelled proudly and simply 'Santorini'.

Today, the island's success with the grape has inspired growers all over the country to try their hand at it, but none ever taste like Santorini. Admittedly, some people find the island's wines a little severe: they aren't fruity trifles, but dry, savoury powerhouses, the sort of white wines that taste best out of a decanter, at a table laden with food, ideally including *domatokeftedes*, tomato fritters that are an island speciality, or even roast lamb.

But once you set foot on Santorini, you'll understand that they couldn't be any other way: these are wines with a vivid sense of place. Note that the island can get very crowded during high season (May–September) – for the best winery experiences, arrive early in the day and book ahead.

GET THERE

Santorini is accessible via 45min direct flights from Athens; a 4hr high-speed ferry runs from Piraeus.

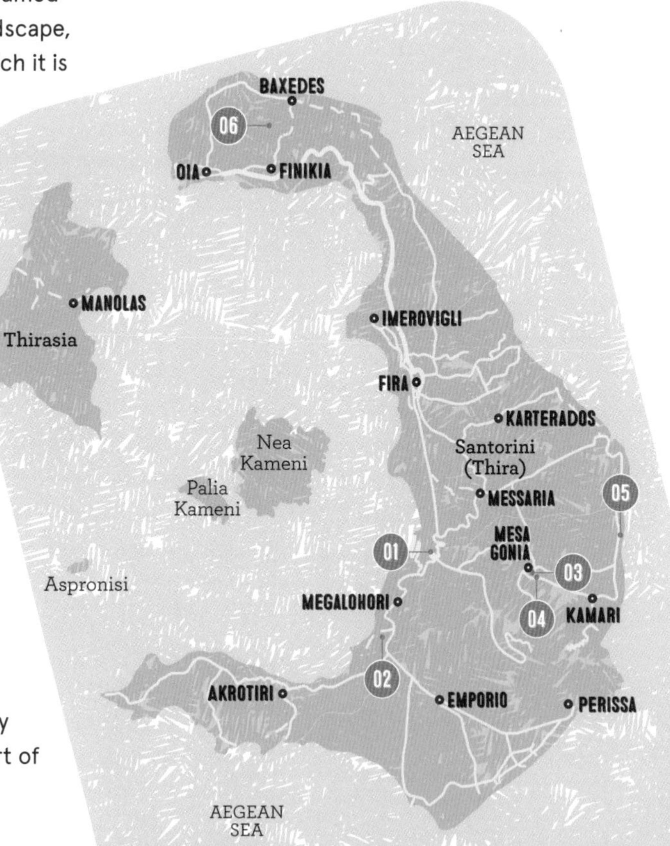

01 Santori coastline

02 Elegant interiors, Estate Argyros

03 Argyros tasting tour

04 Sleek lines at Argyros

01 SANTO WINES

This cooperative functions as Santorini's welcome centre, and plays host to more than a million visitors every year. The location is unbeatable, with the winery perched on a high cliff above the island's main port, offering views over the entire caldera and beyond. It's a terrific spot to get your bearings, as well as to enjoy a taste of the island's gustatory offerings. Winemaker Nikos Varvarigos works with dozens of growers, crafting a vast array of Assyrtiko wines – including the island's only sparkling version – as well as bottlings of rare grapes such as Athiri, Aidani and Mavrotragano. Swing through the gift shop before leaving to stock up with some local produce: the cooperative's members also farm tomatoes and fava, specialities of the island which are protected by PDO (Protected Designation of Origin), as well as capers.

www.santowines.gr; tel 22860 22596; Pyrgos; daily

02 BOUTARI WINERY

The Boutari brand is based in Naoussa, but its influence throughout Santorini can't be overstated. When the company arrived to check out the potential of the island in the mid-1980s, the island's vineyards were rapidly losing ground to tourism construction, and few wineries even carried out any bottling. Boutari's research, work and investment here catalysed the wine scene; it also pioneered the modern style of Santorini by introducing temperature-controlled stainless-steel tanks that allowed the production of fresh, crisp whites. Over time, its wines have also proved Santorini's ageability, with a cellar full of older bottles mellowing into rich complexity. A visit here is a warm, friendly affair, with a walk through the vineyards and traditional mezes on offer if you reserve ahead.

www.boutariwinerysantorini.gr; tel 22860 81011; Megalochori; Mon–Fri in winter, Mon–Sat in summer

03 CANAVA ROUSSOS

Head to Canava Roussos, where the Roussos family has been making wine since 1836, for a special bit of time travel. The *canava* (cellar) is perfectly preserved, with old photos and farming utensils, cisterns and barrels that speak of the island's winemaking traditions; arrive during harvest and the family may even be foot-treading grapes in the stone trough for a special cuvée. The main production now takes place in a modern space, but the winery's speciality is still traditional wine styles, such as Nychteri, a dry, barrel-aged Assyrtiko made from super-ripe grapes, and Mavráthiro, a sweet, spicy wine crafted from a rare red grape. The winery's leafy terrace offers an idyllic place to take in Santorini hospitality.

www.canavaroussos.gr; tel 22860 31278; Episkopi Gonia; daily May–Oct

04 ARGYROS

Growing grapes on Santorini's soil-less, wind-whipped moonscape is an art: because of the strong winds, sun and heat, the locals traditionally train their Assyrtiko vines into *koulouri*, basket shapes nestled close to the ground. And because the island has never been touched by phylloxera, the vine-destroying louse that's forced vintners worldwide to replant their vines on resistant rootstock, Santorini is one of the few places that's home to vines well over 100 years old. Some of those ancient plants grow in a plot just in front of Estate Argyros, the largest landholder on the island. Fourth-generation owner Matthew Argyros is their protector, farming the vines organically and cultivating the oldest plots with donkeys; his best fruit goes into the crisp, saline Estate Santorini and the smoky, rich Cuvée Monsignori, pulled from a patch of 200-year-old vines. He also oversees the island's richest collection of Vin Santo, an elixir made from sun-dried grapes that was world-famous in Venetian times. Incredibly sweet and satiny and yet not at all cloying, these wines can live for decades,

05 Boutari winery

06 Tending the vines

07 Oia

developing bewitching complexity. If you have time, stop at Argyros' newest venture, the micro-sized Volcanic Slopes Vineyard. Housed in a restored 300-year-old winery, it's one of Santorini's most beautiful spaces, dedicated to producing a single wine yearly.
www.estateargyros.com; tel 22860 31489; Episkopi Gonia; daily

05 DOMAINE SIGALAS

Domaine Sigalas sits far away from the crowds, on a gentle slope overlooking the sea on the northeast end of the island. The restaurant alone is reason enough to make the trip; a chic and warm space with exceptionally good food and terraces that extend right into the vineyards. You'll note, however, that some of the vines look like those in other parts of the world, growing vertically on trellises instead of in basket shapes close to the ground. That's because they are Mavrotragano, a variety that owner Paris Sigalas has rescued from near oblivion; he now bottles their fruit in brooding black wines. But try his Assyrtiko wines first, some of Santorini's most elegant, reflecting the gentler environment as well as Paris' Burgundian training. Of special note is his Seven Villages collection, each wine sourced from a different part of the island yet made identically in order to highlight the unique taste of each place.
www.sigalaswinetasting.com; tel 22860 71644; Baxes, Oia; daily

06 GAÍA WINES

Gaía is Santorini's only beachfront winery, an industrial-cool space carved out of what was once a tomato-paste factory. It's run by Yiannis Paraskevopoulos, who is also in charge of a winery in Nemea, in the Peloponnese, and who makes the island's best beer, Crazy Donkey, at the nearby Santorini Brewing Company. A professor of oenology in Athens, he blends a scientific approach with wild imagination – for instance, he's ageing a portion of his wines in metal cages in the cold, dark, oxygen-free depths of the sea. His Thalassitis, on the other hand, is classic, modern Assyrtiko, palate-whettingly bright and acidic; Wild Ferment is a richer, gentler take born out of experiments with wild yeasts. And the grapes that don't make the cut for Gaía's best wines? They are transformed into a memorable aged vinegar that could easily hold its own against Italy's best balsamics.
www.gaiawines.gr; tel 22860 34186; Exo Gonia; daily May–Oct

ESSENTIAL INFORMATION

WHERE TO STAY

THE VASILICOS

The Valambous family have transformed their ancestral summer house into an intimate hotel. With its quirky maze of seven rooms carved out of the volcanic rock, eclectic furnishings and friendly staff, it feels more like a home than a hotel, albeit one with an exceptional chef and wine cellar. *www.thevasilicos.com; tel 22860 23143; Imerovigli*

YOUTH HOSTEL ANNA

It's nearly impossible to find accommodation priced at under €200 a night in Santorini, but this hostel in Perissa fits the bill. It's pretty basic, but offers free pickup from the airport, air-con, a pool and private rooms. What it lacks in views it makes up for in convenience: black-sand beaches are in easy walking distance, and it's far removed from the tourist crush of Fira. *www.hostelannasantorini.com; tel 22860 85401; Perissa*

07

WHERE TO EAT

SELENE

Chef Yiorgos Hatziyannakis was farm-to-table long before it became hip, opening this restaurant dedicated to local food in 1986. The superb cuisine is complemented by Santorini's best wine list, with magnificent sunset views a bonus. Below the restaurant, his casual cafe offers more traditional meze, like silken fava purée with grilled cuttlefish and caper leaves, or saffron-scented seafood stew. *www.selene.gr; tel 22860 22249 (restaurant), tel 22860 24395 (cafe); Pyrgos*

TO PSARAKI

High above the Vlychada marina, To Psaraki offers super-fresh fish paired with an excellent array of Santorini wines. Fish are priced by the kilo; talk to the waiter to make your choice and discuss preparation. Chances are, it will be grilled, needing nothing but a drizzle of *ladolemono* – a creamy emulsion of lemon juice and olive oil – and a glass of Assyrtiko to be out of this world. *www.topsaraki.gr; tel 22860 82783; Vlychada*

WHAT TO DO

WALKING

One of the island's many walking trails starts in Fira and traces the cliff edge out to the promontory of Skaros Rock, with its ancient Venetian ruins, and winds up in northerly Oia. Another climbs to the top of Pyrgos, at 2000m (6562ft), twisting through vineyards and wild fava, caper bushes and saffron poppies, on its way to the 18th-century monastery of Profitis Ilias.

ARCHAEOLOGICAL SITE OF AKROTIRI

Often referred to as the Greek Pompeii, Akrotiri was a thriving city until the catastrophic Minoan eruption encased it in pumice. Excavations reveal an extensive and dazzlingly well-preserved city, its buildings and colourful frescoes offering a vivid glimpse into Santorini's illustrious history.

CELEBRATIONS

Most months feature feast days, religious holidays celebrating certain saints. Held in Emporio on 22 October, the most wine-soaked affair is the feast of Agios Averkios, patron of wine.

01

[Hungary]

TOKAJ

Take a trip back in time, with vivacious and friendly companions, on a journey through Hungary to taste the mystical and ancient Tokaj wines.

HUNGARY

SLOVAKIA

SÁTORALJAÚJHELY

HUNGARY

SÁROSPATAK

Bodrog River

Tisza River

TOKAJ

01 02 03 04 05 06

Tisza River

Picture a chilly autumn dawn. Along the low slopes of celery-green hills in northeastern Hungary, the moist air sends up mists that kiss the vineyards of an undulating, Unesco-protected landscape. For many grapes growing here, the peculiar botrytisation process has begun. Already deliberately over-ripened, the fruit starts rotting, and in doing so it drastically sweetens and acidifies the grapes to form the singular taste of the legendary Tokaj Aszú wine.

Aszú was never short on admirers. Austro-Hungarian Empress Maria Theresa and Napoleon sang its praises. Louis XIV dubbed it 'wine of kings, king of wines'. Amber-hued, sweeter and more intense than any other wine in the world, the ultimate maverick of top-quality vintages is every bit as surprising to first-timers today. Aszú is the internationally recognised hallmark of the region, but it's only the kick-off for the oenophile's odyssey. Despite a pedigree going back to the 16th century, a half-century of communist practices dented Tokaj's prestige, obliging the industry to reinvent itself post-1989. First it happened out on the terraces and in the century-old cellars: winemakers discovered an improbably diverse range of dry whites could flourish in Tokaj too, and these proved a hit with connoisseurs. Then entrepreneurs realised Tokaj could tout more than just wineries, opening wine bars, wine spas, wine hotels and so on.

Because Tokaj is still playing catch-up, it has an undiscovered, utterly unpretentious feel. Winemakers happily chat to visitors. And, as this is a region large on character, small on production and with little presence in international markets – a trip here guarantees a sip into uncharted territory.

GET THERE

Budapest Ferenc Liszt is the nearest major airport, 225km (140 miles) from Mád. Car hire is available.

01 Tokaj Kikelet Pince vineyards

02 Patricius cellars

03 Vines at Disznókő

04 Tasting at Tokaj Kikelet Pince

01 HOLDVÖLGY

No one's saying the wine quality at Holdvölgy is anything but the finest. But the 2km (1.2-mile) network of cellars is the highlight here, and makes this a great family choice. Plumb in the midst of the twisting streets of central Mád, this place only opened to the public in 2013.

Before the cellar descent, visitors are furnished with a map of the passages and then they're on their own (well, a guide will bring up the rear in case anyone gets lost in the labyrinth). The challenge is to find where the wine is stashed, treasure-hunt style, and return to the surface – if you can!

www.holdvolgy.com; tel 70 391 4643; Batthyány utca 69, Mád; by appointment

02 DISZNÓKŐ

Even the tractor garage at photogenic Disznókő is a work of art (the startling design is based on a yurt and a volcano, apparently). The region's prettiest winery seems to excel at everything. It hosts a monthly Sunday produce market that's a networking event for winemakers; it has a highly regarded restaurant and a wine shop, and it operates as a wedding venue. Visitors are also encouraged to take a turn around the estate: the 19th-century belvedere has some fabulous views.

But where the wine is concerned, Disznókő is intensively focused. As one of the bigger Tokaj winemaking enterprises, it has the wherewithal to concentrate exclusively on producing the expensive, time-consuming Aszú wines. In this respect, Disznókő is groundbreaking, and the vintages it turned out in 1993 represented the first of the striking new, post-communist Aszú. Gone was the old-school heaviness, and in its place came citrusy freshness, along with a swing towards premium five- and six-*puttonyos* wines (the more *puttonyos*, the higher the sugar content). Today, Disznókő remains the most active innovator in the Tokaj business, and it's well worth taking time to visit the winery.

www.disznoko.hu; tel 47 569 410; near Mezőzombor; by appointment

04

03 GRÓF DEGENFELD

Kudos to Gróf Degenfeld for being one of the few eco-wineries. The majesty of its lush landscaped grounds, elegant hotel and character-rich winery is enormously augmented by pesticide-free vineyards that lend a timeless pastoral feel to the estate rearing up behind its luxury Castle Hotel. Every second row is planted with wildflowers to attract the bugs that kill harmful pests. Grapes are hand-picked. It even has its own forest to grow the wood for its barrels.

The Degenfelds are one of Hungary's old families, and the 200-year-old cellars, cutting back deep into the hill, still retain their 19th-century charm, including the original Degenfeld coat of arms on the creaking oak doors. On the varied wine list, look out for the signature wine: a late-harvest partially botrytised Furmint. There is also a good wine shop on-site.
www.grofdegenfeld.hu; tel 47 380 173; Terézia kert 9, Tarcal; by appointment

04 TOKAJ KIKELET PINCE

As with many things wine-related, the French played a part in Hungarian Tokaj's brand reinvention post-1989. Stéphanie Berecz, hailing from the Loire, exemplifies this trait. One of the region's best-regarded small producers, she refined her trade overseeing operations at nearby Disznókő, and the dinky winery that she and her husband now run is the antithesis of the corporate experience: extremely personal and, by her own admission, utterly unpredictable.

'Every year the wine is different,' says Berecz. 'Generalisations can never be made. A priority for us is being authentic: always faithful to the soil and the climate and their conditions. This belief is why we have developed our own yeasts, which is to do with being natural, and wanting to make the wines most representative of the area.'

Surprise tastings aside, tours include a hike above the winery's vineyards to a lake in a former quarry – good for a swim, but also to get an idea of the regional rock

types that can influence the taste of the wine they produce.
www.tokajkikelet.hu; tel 30 636 9046; Könyves Kálmán utca, Tarcal; by appointment

05 ZOLTÁN DEMETER PINCÉSZET

Unfortunately for the wine region's namesake town, Tokaj itself has few top-notch wineries – but the one run by Zoltán Demeter compensates for all that. This one-person, 7-hectare (17-acre) operation appeals mostly because of the charming idiosyncrasies of the owner. Wines here, for instance, mature to the sound of Mozart. Demeter makes wine across the three categories he deems most defining of the Tokaj region, namely Aszú, Főbor (also known as *szamorodni*, or principal wine, another traditional Tokaj wine now fast becoming as popular as Aszú) and dry white.
www.facebook.com/demeterzoltantokaj; tel 20 806 0000; Vasvári Pál utca 3, Tokaj; by appointment

06 PATRICIUS

The vineyards of Tokaj-Hegyalja were the first to be classified (over a century before Bordeaux, for example): designated as first-, second- or third-growth as early as 1720. First-growth vineyards don't guarantee first-class wines but Patricius claim to have eight, which certainly helps. These guys are major Tokaj players, but their pearl-white winery feels anything but frenetic, ensconced as it is in the middle of the vineyards.

The remains of a castle crown the hill above; beneath, in a plot laid out the old-fashioned way (as opposed to machine-friendly rows of vines), is a living museum exhibiting grape varieties present and past, including several the phylloxera epidemic eradicated. Despite the 19th-century buildings, the winery itself is modern, with a gravity-flow system (meaning wine is produced over two levels and takes advantage of gravity to move around the site more gently, minimising the chance of the flavour being compromised) and a state-of-the-art tasting room, which doubles as a gallery.
www.patricius.hu; tel 47 396 001; near Bodrogkisfalud; daily

05 Picnic at the Patricius vineyards

06 Patricius wines

ESSENTIAL INFORMATION

WHERE TO STAY

BARTA PINCE WINERY
A graceful guesthouse with huge rooms in a 16th-century farmhouse. The adjacent winery, which has steep-sloping vineyards gaining maximum flavour from the elements, already has a reputation. The owners also have a corker of a mansion, the Rákóczi-Aspremont, with three palatial suites available. *www.bartapince.com; tel 30 324 2521; Rákóczi utca 83–85, Mád*

GRÓF DEGENFELD CASTLE HOTEL
This delightful terracotta-roofed mansion sits within landscaped grounds at the foot of the namesake winery. The style is Hungarian Empire heyday; tennis courts and a pool are alongside. *www.hotelgrofdegenfeld.hu; tel 47 580 400; Terézia kert 9, Tarcal*

WHERE TO EAT

ELSŐ MÁDI BORHÁZ
This vineyard-facing bistro showcases the majority of wines in the Mád Circle (Mád winemakers association) with a cutting-edge wine dispenser. It will pack you picnics to take to the vineyards. *www.mad.info.hu/elso-madi-borhaz; tel 47 348 007; Hunyadi utca 2, Mád*

PERCZE ÉLÉMENYKÖZPONT
Tokaj's most famous ambassador, István Szepsy, no longer offers visits to his winery, but this restaurant, opened by his granddaughter in 2016 and situated on a vineyard, has all the range of Szepsy wines, plus exquisite food and attentive service. *Tel 20 464 2222; Árpád utca 70, Mád*

GUSTEAU
Refined high-end dining (or 'culinary experience workshop', to use its own phraseology) in a tucked-away courtyard restaurant, focusing on serving food to match the area's phenomenal wines. The wine dinners offer several courses, each of which is paired with a different regional vintage. *www.gusteaumuhely.com; tel 47 348 297; Batthyány utca 51, Mád*

WHAT TO DO

TOKAJ TODAY TOURS
You'd be hard pressed to say whether it's Greg Somogyi's insightful knowledge of Tokaj or his impeccable English that impresses most; either way, he'll take you on some memorable jaunts into the heart and soul of the wine region. *https://travel.tokajtoday.com*

ÍZES ÖRÖMEST
Ease off the booze at this husband-and-wife jam making outfit: he manages the winery and she makes the wine-inspired preserves. *www.facebook.com/izesoromest, Bodrogkereszturi ut 48b, Tokaj*

VINOSENSE SPA
The spa at Andrassy Residence in Tarcal offers Aszú facials and body wraps. *andrassyrezidencia.hu*

CELEBRATIONS

For one Saturday in June, Good Night Mád sees cellars and restaurants in Mád stay open late into the night for music, feasting and wine-fuelled frivolity. Come July, Wine, Shine, Bénye brings together winemakers and bands over three days in a similarly vintage event. *www.tokajtoday.com*

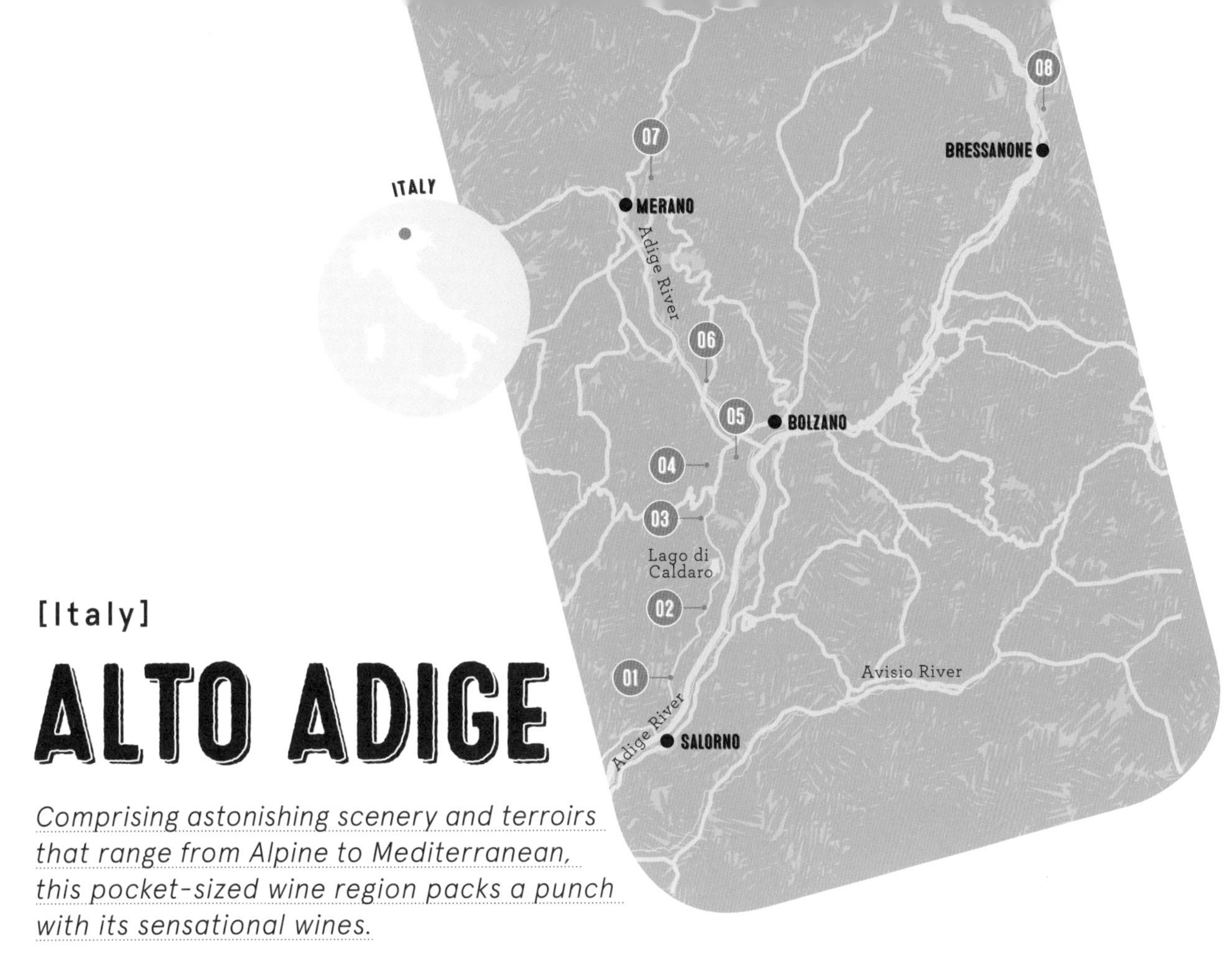

[Italy]

ALTO ADIGE

Comprising astonishing scenery and terroirs that range from Alpine to Mediterranean, this pocket-sized wine region packs a punch with its sensational wines.

The Italian Tyrol state of Alto Adige stretches up as far as the Alpine frontier with Austria, holding a unique semi-autonomous status. Village names are expressed in both Italian and German – from Bolzano (Bozen) in the south up to Bressanone (Brixen) in the north. Even though the region begins roughly just an hour up the *autostrada* from Verona, locals here will greet you with a cheery *Grüss Gott* instead of ciao, and rather than pasta and tiramisu, the favourite dishes are *canederli* bread dumplings and apple strudel.

The native Rhaetian tribes of Alto Adige were making wine and storing it in barrels long before the Romans arrived but, until recently, cooperative cantinas opted for low-quality, bulk production. Not today, though. You will find some of Italy's finest whites here: Pinot Bianco, Pinot Grigio, Chardonnay, Sauvignon and Gewurztraminer. Two indigenous reds deserve to be better known: the light, drinkable Vernatsch and intense Lagrein, and some cuvées of Pinot Nero Riserva can rival even those of Burgundy.

The scenery on Alto Adige's historic wine trail (*Strada del Vino*) is outrageously beautiful. It passes through the base of steep glacial valleys planted with thousands of fruit trees, while both sides are covered by a geometric maze of criss-crossing vineyards. This is one of the oldest wine trails in Italy, started back in 1964, and it comprehensively covers the whole of the region. No matter how lost wine enthusiasts may get, there is always the distinctive sign of the *Strada del Vino* pointing them in the right direction. Throughout the year, the association organises walking, biking and horse-riding trips through the vineyards, as well as gastronomic celebrations in winemaker villages.

GET THERE
Venice Marco Polo is the nearest major airport, 250km (155 miles) from Magrè. Car hire is available.

01 Alpine scenery in Alto Adige

02 Cantina Tramin winery

03 A vegetarian dish served at Alois Lageder

04 Tasting at Paradeis manor, Alois Lageder winery

01 ALOIS LAGEDER

Driving into Alto Adige from the neighbouring Trentino region, the first obligatory stop-off is the cantina of Alois Lageder, a pioneer winemaker. Way back in 1995 Alois created a revolutionary cantina, powered by solar energy, today, the family's own 55 hectares (135 acres) are both organic and biodynamic, while over 50% of the 80 smallholders they buy grapes from have converted too. Tastings are held at Paradeis, a rambling 15th-century manor, with a small restaurant serving mainly vegetarian organic cuisine. It is quite a task deciding what to select from the extensive list of 35 different wines but be sure to ask for the Lowengang, an impeccable Chardonnay, and Lagrein Lindenburg, a fruity red made from a native grape.

www.aloislageder.eu; tel 0471 809 580; Via Casòn Hirschprunn 1, Magrè; Mon–Sat

02 CANTINA TRAMIN

The sign at the entrance to the enchanting village of Tramin proudly states that this is 'home' of the world's Gewurztraminers, the aromatic grape that has been grown here for a thousand years and is now found all over the globe. At the outskirts of town right on the *Strada del Vino*, you can't miss the dramatic avant-garde winery of the *cantina sociale* (cooperative association), resembling a maze-like green cube with a panoramic glass tasting room. The cantina sold most of its wine *sfuso* (in bulk) until Willi Sturtz was appointed winemaker in 1992. He began to raise the quality in the vineyards and cut yields. The cantina now produces 1.8 million bottles of high-quality wines annually. While both the Pinot Grigio and Müller-Thurgau are surprising, the special Gewurztraminer Nussbaumer Cru is simply exceptional.

www.cantinatramin.it; tel 0471 096 633; Strada del Vino 144, Tramin; Mon–Sat

03 KLOSTERHOF

Guests arriving at the *weingut* (estate) winery of Oskar Andergassen can be sure of a warm welcome from this jolly

04

vignaiolo (winegrower). He is one of the new generation of small producers who have stopped selling grapes, choosing instead to make his own vintages from a perfectly positioned 4-hectare (10-acre) vineyard situated high above Lake Caldaro. Oskar concentrates on Pinot Bianco, Vernatsch and Pinot Nero, all very distinctive. The Pinot Bianco is aged in acacia-wood barrels, while the Vernatsch comes from ancient vines cultivated in the traditional pergola system. Klosterhof is not just a cantina but a comfortable Tyrolean hotel too, with a vineyard pool and a cosy *weinstube* (wine bar), the perfect place to try home-cured speck, ham and delicious mountain cheeses.

www.klosterhof.it; tel 0471 961 046; Prey-Klavenz 40, Caldaro; daily by appointment

04 CANTINA COLTERENZIO

Wherever you are in Alto Adige, it's impossible to miss the distinctive black tower that marks the wines of the Colterenzio's *cantina sociale*. At any time of day there's a steady stream of wine enthusiasts filling the tasting room of this contemporary winery, which has a range that stretches from affordable, sharp Chardonnay and Pinot Grigio through to vintages such as Cornelius, a tremendous Merlot Cabernet blend honoured by the coveted Tre Bicchieri, Italy's top wine award. The *cooperativa* was founded in 1960 by just 28 small vintners; today there are 300 *soci* (members).

www.colterenzio.it; tel 0471 664 246; Strada del Vino 8, Cornaiano; Mon–Sat

05 STROBLHOF

The date 1664 is carved into the entrance of Stroblhof, and *vignaiolo* Andreas Nicolussi explains, 'Wine has always been produced at our *maso*, to go with the meat produce of the farm. A plate of speck is not complete without a glass of Vernatsch – this is what everyday life is all about here.' Today, guests staying at Stroblhof have a much more luxurious time, spoilt by a chef and pampered with a wellness spa and freshwater lake pool.

The soil here is a mixture of volcanic and chalk, which produces a sharply acidic, mineral Pinot Bianco – which Andreas ages in large barrels – as well as a tannic Pinot Nero that stays 18 months in small *barriques*.

www.stroblhof.it; tel 0471 662 250; Via Pigano 25, Appiano; daily

06 CANTINA TERLANO

Terlano is the story of a far-sighted winemaker, Sebastian Stocker, who transformed this relatively small, conservative *cooperativa* into quite possibly

05 Cloisters at Abbazia di Novacella

06 Novacella wines

07 Abbazia di Novacella collegiate church

08 Autumn in Alto Adige

the leading cantina in all of Alto Adige. Stocker was convinced that Terlan's microclimate – hot days, cool nights – combined with the porphyr volcanic soil was perfect for ageing white wines. For many years the *soci* would have none of this, preferring to sell the wine young. So every year, beginning back in 1955, Stocker stashed a couple of hundred bottles of each wine in the cellar's hidden nooks and crannies. When he finally revealed all, the *soci* were furious, until they tasted how wonderfully the wines had aged.

Stocker's successor, Rudi Kofler, has continued to make award-winning yet affordable wines. Be sure to book ahead for a visit to the futuristic cellar.

www.cantina-terlano.com; tel 0471 257 135; Via Silberleiten 7, Terlano; Mon-Sat

07 INNERLEITER

The narrow lane up to Innerleiter zigzags through a series of hair-raising bends, emerging at a romantic *gasthaus* hotel with breathtaking views across the snowcapped Alps. Karl Pichler cultivates just 1.7 hectares (4 acres) of vines encircling the hotel. He has built a modern cantina and works alone without a wine consultant, making some bold decisions, such as to totally abandon cork in favour of screw-top. 'Many hotels here have vineyards like us,' he says, 'but they are always separate, and I wanted to bring the cantina and hotel together. So you can sit down in our tasting room, try the chef's pairing snacks with each glass, and see the barrel room through a glass wall.'

www.innerleiterhof.it; tel 0473 946 000; Via Leiter 8, Scena; Wed-Mon

08 ABBAZIA DI NOVACELLA

Even if this was not the home of some of the finest wines in Alto Adige, it would still be worth the detour to discover this magnificent Augustinian monastery, with its baroque chapels and ornate library. It has been cultivating vines since its foundation in 1142. The Abbazia lies just outside Bressanone, and its white-wine vineyards rise up to almost 1000m (3280ft). Red wines come from another monastery near Bolzano, but in the Abbazia be sure to taste fruity, mineral Sylvaner, Müller-Thurgau, Veltliner and the full-bodied Kerner, a curious cross of Riesling and Vernatsch. And don't miss the aromatic Moscato Rosa, which could easily be called rose-petal wine.

www.abbazianovacella.it; tel 0472 836 189; Via Abbazia 1, Varna; Mon-Sat

ESSENTIAL INFORMATION

WHERE TO STAY

GASTHAUS KRAIDLHOF
This tiny farmhouse surrounded by orchards and vines has a pool disguised as a freshwater pond in a Zen garden. *www.kraidlhof.com; tel 0471 880 258; Hofstatt 2, Kurtatsch*

DER WEINMESSER
Christian Kohlgruber is a wine fanatic and sommelier. His luxurious hotel revolves around wine, from trips to the vineyard to tastings in the cellar and vinotherapy in the spa. *www.weinmesser.com; tel 0473 945 660; Via Scena 41, Scena*

GASTHOF HALLER
At the edge of bustling Bressanone, this Alpine chalet is a haven of peace where you can almost touch the owner's vineyard from your window. Food is served in a snug wood-panelled *weinstube*. *www.gasthof-haller.com; tel 0472 834 601; Via dei Vigneti 68, Bressanone*

WHERE TO EAT

PFEFFERLECHNER
This unique cantina has a dining room that looks directly into a stable with horses and cows. There's a beer garden, a microbrewery and a copper alembic that's used to distil grappa. *www.pfefferlechner.com; tel 0473 562 521; Via San Martino 4, Lana*

HOPFEN & CO
In the heart of medieval Bolzano, this 150-year-old osteria is the perfect place to sample such classic Tyrolean dishes as roast pig's knuckle and sauerkraut. *www.boznerbier.it; tel 0471 300 788; Piazza Erbe 17, Bolzano*

WHAT TO DO

Spend a day at Merano's historic Thermal Baths (Terme Merano) with its pools, sauna and spa. *www.termemerano.it; tel 0473 252 000; Piazza Terme 9, Merano*

CELEBRATIONS
Merano's wine and food festival is held in November; Egna hosts Pinot Nero Days in May, and in April/May the annual asparagus season is celebrated with wine pairings in Terlan. *Törggelen* season in cantinas begins at the end of September, when you can taste partly fermented grape juice that's served alongside freshly roasted sweet chestnuts.

07

Ø8

[Italy]

FRIULI

In undiscovered northeast Italy, snowy mountains and fertile plains are reflected in the variety of wines, from intense reds to fragrantly sweet whites.

The rugged Friuli region stretches from the shores of the Adriatic up to the Alps, forming a wedge between Italy's border with Eastern and Central Europe. Vineyards spread along the flat plains of the Piave, Italy's 'sacred river', where rough Raboso wine was a great favourite of Ernest Hemingway, to the Carso, a rocky peninsula running up towards Trieste, where cantinas are often hewn into underground caves.

Inland, the Collio Orientale – the eastern hills around the ancient town of Cividale – are famous for fascinating reds with such intense local grapes as Refosco and Pignolo, but the jewel in Friuli's crown is the Collio, a 50km (31 mile) necklace of hills.

The clay and sandstone soil here produce some of the finest white wines in Italy: fruity indigenous grapes such as Ribolla Gialla; the unique Picolit, late-harvested for a luscious dessert wine rivalling Sauternes; and the local favourite, Friulano, once known as Tokai, though today this name can only be used by Hungary's famed sweet wine.

Many Collio winegrowers have opened up their estates as B&Bs, often inviting guests to whizz around the vineyards on signature bright-yellow Collio Vespas, and as Friuli is still very much undiscovered, you are sure of a warm welcome. The same is true of eating out; all over the countryside there are rustic *agriturismi* (farmstays) that open at the weekend and offer traditional Friulian fare, which is more influenced by Central European cuisine than Italian. Plump gnocchi stuffed with susina plums is perfect with a sharp Friulano, and the more characteristic Ribolla Gialla goes well with juicy baby squid sautéed with slightly bitter red radicchio. A popular dish is Friuli's rich goulash stew; it's worth opening a bottle of one of the region's stellar reds, Livio Felluga's Sossó, a potent combination of Merlot and Refosco, to accompany it.

GET THERE

Trieste-Friuli Venezia Giulia is the nearest major airport, 30km (19 miles) from Dolegna del Collio. Car hire is available.

01 VENICA & VENICA

Just before you drive into the sleepy village of Dolegna, a small sign on the right directs you down a narrow route that leads to one of the Collio's most important wineries. Venica & Venica refers to two brothers, Gianni and Giorgio, who have turned the small vineyard founded by their grandfather 80 years ago into a slick, modern estate spanning 37 hectares (90 acres). The one wine not to miss here is Ronco Bernizza, a surprising, steely Chardonnay that's perfect with *spaghetti alle vongole* (spaghetti with baby clams).

www.venica.it; tel 0481 61264; Località Cerò 8, Dolegna del Collio; Mon–Sat

02 CRASTIN

Crastin is the tiniest hamlet imaginable, with a single ancient farmhouse where Sergio Collarig cultivates a small 7-hectare (17-acre) property. He is a rough-and-ready *contadino*, what might romantically be termed a peasant farmer, aided by his sister Vilma. Together they have progressed from producing *vino sfuso* (wine sold in bulk) to creating a small garage cellar yielding not just the Friulano and Ribolla Gialla whites that Collio is so famous for, but also full-bodied Merlot and Cabernet Franc aged in oak barrels. Each weekend they open up as an *agriturismo*, with Vilma preparing generous plates of ham, sausages and cheeses while Sergio opens bottles for tastings.

www.vinicrastin.it; tel 0481 630 310; Località Crastin 2, Ruttars; Sat–Sun & by appointment

03 LIVIO FELLUGA

This historic family winery sets the benchmark for excellence in both the Collio and adjoining Collio Orientale vineyards, its vast estate covering a total of 160 hectares (395 acres). The founding father of the estate, Livio Felluga, who passed away in 2016 at the age of 102, declared that, 'There were many doubters when I started clearing forest land and planting vines here 60 years ago, but history told me that wine had been produced here for centuries and I was sure that this was the perfect place for white grapes

01 Autumnal Friuli vineyards

02 Picking crew, Venica & Venica

03 Ripening on the vine

04 Livio Felluga estate

'I grow wonderful grapes. I just do the minimal fine-tuning in my cantina so as not to spoil the grapes and let them make their own magic.'

– Paolo Caccese, winemaker

like Friulano, Sauvignon and Pinot Grigio and our indigenous red Refosco.' The family recently took over the Abbazia di Rosazzo, a magnificent frescoed abbey surrounded by vineyards, where monks began making wine a thousand years ago; it now serves as the venue for wine tastings.

www.liviofelluga.it; tel 0432 759 091; Abbazia di Rosazzo, Piazza Abbazia 5, Località Rosazzo, Manzano; Mon–Sat $

04 PAOLO CACCESE

The hamlet of Pradis stretches over a series of rolling vine-clad hills whose sole inhabitants are a dozen *viticoltori*, who all make exceptional wines. Paolo Caccese's cantina sits atop the highest hill, surrounded by his 6-hectare (15-acre) vineyard. Paolo is a genuine eccentric, dressed like a country gentleman, and resembling more the lawyer that he trained to be than a producer of a clutch of elegant wines. His classic Friulano and Malvasia are delicious, but ask to try such oddities as the fruity Müller-Thurgau, aromatic Traminer and a luscious late-harvest Verduzzo. He defiantly ignores trends and fashions, still uses old-fashioned cement vats, and explains, 'I grow wonderful grapes here on a rich soil in an incredible position on the hillside, so I just do the minimal fine-tuning in my cantina so as not to spoil the grapes and let them make their own magic.'

www.paolocaccese.it; tel 348 797 2993; Località Pradis 6, Cormòns; daily by appointment

05 RENATO KEBER

The meandering road that leads out of Cormòns towards San Floriano is marked on both

05 A tasting at Venica & Venica

06 The Abbazia di Rosazzo at Livio Felluga

sides by the Collio's distinctive winemaker signs, and at Zegla, a narrow lane leads you to the domain of Renato Keber, a one-of-kind *vignaiolo*. A quiet and unassuming man, Renato has built a swanky tasting room that affords panoramic views over his vineyards, where he loves surprising visitors with his spectacular wines.

Renato waits seven years before bringing out each of his Merlot and Cabernet vintages, but also follows the same philosophy with his whites – so it can come as quite a shock when he opens, say, a 2008 Pinot Grigio or a 2007 Sauvignon. 'My wines are marathon wines,' he jokes, 'and it is best to wait at least until you get to the 20km mark!'
www.renatokeber.com; tel 0481 639 844; Località Zegla 15, Cormòns; daily by appointment

06 AZIENDA AGRICOLA FRANCO TERPIN

Franco Terpin is an anti-establishment artisan winemaker, the guru of a small group of natural, no-sulphite wine producers. Certified organic, and favouring long maceration, natural yeast and absolutely no chemicals, Franco's wine spends a year in the barrel, another year in steel vats, then three years ageing in the bottle. He produces 90% white wines on the small estate, which includes vines across the border in Slovenia. They have a quite incredible orange colour, known here as *vini arancioni*.

Franco tells his visitors: 'My wines are natural and, quite frankly, they are the only kind of wines I drink now – I can't stand a Chardonnay that has a banana aroma or the classic cat's pee of Sauvignon – these are chemically induced. With my wines you can drink a few bottles, party till 3am, and have absolutely no hangover the next morning.'
www.francoterpin.com; tel 0481 884 215; Località Valerisce 6/A, San Floriano del Collio; daily by appointment

07 GRADIS'CIUTTA

The Princic family have been cultivating grapes in the Collio and in nearby Slovenia since 1780, though today's modern winery was officially founded in 1997 when winemaker Robert Princic persuaded his parents, Zorko and Ivanka, to expand their working farm to include a serious vineyard. For the last six years, Robert has served as President of the Consorzio, representing all of the Collio winemakers.

His own crisp, mineral wines include a still and sparkling Ribolla Gialla, while his Chardonnay is one of the finest in the region. Visitors are assured of a warm welcome, whether from Robert, explaining all the technical aspects of his barrel-ageing, or Mamma Ivanka slicing up thick chunks of salami produced on the farm.
www.gradisciutta.eu; tel 0481 390 237; Località Giasbana 32/A, San Floriano del Collio; Mon–Sat

ESSENTIAL INFORMATION

WHERE TO STAY

BORGO SAN DANIELE

Mauro and Alessandra Mauri make outstanding wines and have created a designer B&B with a pool adjacent to their cantina. *www.borgosandaniele.it; tel 0481 60552; Via San Daniele 28, Cormòns*

AZIENDA AGRICOLA PICECH

Roberto Picech rents out three rooms in his farmhouse winery, with stunning terraces overlooking vine-blanketed hills. Breakfast is served in the family's own kitchen where wine tastings are held in the evening. Free bikes are provided, too. *www.picech.it; tel 0481 60347; Località Pradis 11, Cormòns*

06

WHERE TO EAT

TRATTORIA AL CACCIATORE DELLA SUBIDA

This once-rustic trattoria has been transformed by the Sirk family into an elegant Michelin-starred restaurant offering dishes such as wild deer carpaccio with fresh horseradish. *www.lasubida.it; tel 0481 60531; Via Subida 52, Cormòns*

AGRITURISMO STEKAR

A working farm and vineyard that Sonia Stekar opens as a restaurant every weekend (and Wed-Fri for dinner, but call ahead to check), serving up goulash, award-winning salamis, pumpkin gnocchi and out-of-this-world apple strudel. *www.bio-stekar.it; tel 0481 391 929; Località Giasbana 25, San Floriano del Collio*

ENOTECA REGIONALE DI CORMÒNS

This pulsating wine bar gives you a chance to mingle with local *vignerons*, sample some of the 100 wines available by the glass and gnaw on generous plates of meats and cheeses. *www.enotecacormons.it; tel 0481 630 371; Piazza XXIV Maggio 21, Cormòns*

WHAT TO DO

From the ancient Roman town of Cividale del Friuli, you can explore the wooded valley along the fast-flowing Natisone river, perfect for hiking, mountain biking and trout-fishing, and not short of appealingly rustic trattorias, either.

CELEBRATIONS

Each September, Gorizia celebrates Gusti di Frontiera, a food and wine festival, while at the end of October the Jazz & Wine festival has a roster of musicians performing in cellars in and around Cormòns.

[Italy]

LIGURIA

Wedged between the mountains and the sea, Liguria is one of Italy's most underrated wine regions, a place of 'heroic winemakers', cliffside vineyards and bold vintages.

Famed for its elegant beaches, pastel-tinted villages and decadent seafood, Liguria has long captivated visitors. Days here are spent gazing out from faded villas at turquoise waters, or walking rocky paths above ancient olive groves and vineyards first planted by the Romans. The wines have long played second fiddle to the region's more obvious charms: the grandeur of upscale seaside towns like San Remo, the candlelit dining rooms overlooking Portofino or the medieval lanes and sumptuous palazzi of Genoa.

In fact, until a few years ago, most Italians would shrug when asked about the attributes and virtues of a Ligurian wine. 'Reds a bit on the fizzy side,' a waiter might have advised – unkindly. 'How about something from Tuscany instead?' But things have changed dramatically in recent years in this crescent-shaped region anchoring Italy's northwest coast. A new generation of winemakers are introducing bold techniques while embracing a sustainable, back-to-the-land approach.

Italians often speak of *viticoltura eroica* (heroic winemaking) when describing the challenges of growing grapes in an extreme environment. No place seems to embody this more than Liguria. Working tiny plots on precipitously steep hillsides, protected by drystone terraces painstakingly built by hand – with heavy storms and landslides a perennial concern – discourages all but the most tenacious of winemakers. One taste of a Vermentino from Colli di Luni or a Rossese from Dolceaqua, and you'll quickly realise why they go to all the trouble. Ligurian wines express a deep-rooted terroir: flavours can evoke Mediterranean herbs, forest berries, and the faint brininess of the salt-kissed air.

GET THERE
Genoa, on the coast near the centre of Liguria, has a few international flights, with more options from Milan (185km/115 miles north of Genoa).

ITALY
CUNEO
GENOA
BORZONASCA
PONTREMOLI
RAPALLO
SAVONA
Gulf of Genoa
SESTRI LEVANTE
Riviera di Levante
FIVIZZANO
FINALE LIGURE
05
02
03
01
FRANCE
Riviera di Ponente
LA SPEZIA
MASSA
ALASSIO
IMPERIA
04
Ligurian Sea
06
SAN REMO
MONACO
MEDITERRANEAN SEA

01 BURANCO

A short stroll from the village of Monterosso al Mare, Buranco produces some fine DOC Cinque Terre white wines as well as grappa, *limoncino* (Liguria's answer to the Amalfi Coast's lemon-flavoured limoncello) and one tricky-to-pronounce dessert wine: *Sciacchetrà* (shah-keh-trah). The amber-yellow, aromatic wine has an intense nose and is richly complex with notes of dried fig, candied orange and hazelnut. It's quite labour-intensive to make: meticulously selected grapes are hand-picked then dried naturally indoors on trellises for two months before fermenting, followed by 16 months ageing in steel barrels. Before or after a tasting on the veranda, you can wander the vineyards to get a sense of the challenges of working the steeply terraced hillsides. Buranco rents out three apartments on the property, all with terraces (two overlooking the vineyards). It also runs a small restaurant, specialising in locavore cuisine, in Monterosso's old town.
www.burancocinqueterre.it; tel 0187 817 677; Via Buranco 72, Monterosso; daily by appointment

02 CANTINA CINQUE TERRE

High above the seaside village of Manarola, the Cinque Terre's largest wine producer has been cultivating vines since 1982. Over 200 grower-members make up this cooperative, which consists of dozens of lofty parcels of land, spread across the southern-facing hillsides above the crashing waves. Some of the best wines here hail from historical vineyards that date back centuries. Sipping the DOC Costa de Campu, you can almost taste the summer heat and Mediterranean wildflowers in this golden-hued wine with its notes of sage and citrus, while the Costa de Posa evokes the constant sea breezes with a delightfully briny mineral finish. This is also the place to find one of Liguria's best *Sciacchetràs* – an intense award-winning Riserva that's aged in small oak barrels for at least three years.
www.cantinacinqueterre.com; tel 0187 920 435; Località Groppo, Riomaggiore; daily by appointment, closed Sun Jan–Mar & Nov

01 Cinque Terre coastline at Manarola

02 Liguria's mountainous interior

03 Tasting Dolceacqua at Terre Bianche

03 POGGIO DEI GORLERI

In a tiny settlement above the Riviera del Ponente (the western Riviera, which runs from Genoa to the French border), the Merano family founded this winery back in 2003, determined to create a great Ligurian wine that marries a commitment to innovation with a respect for tradition. The south-facing slopes overlooking the sparkling Gulf of Diano Marino make a rugged setting for growing, though the iron-rich and limestone soils provide the foundation for excellent Vermentino and Liguria's native Pigato grapes. Their Albium, made of 100% Pigato, has a warm and bright character that's remarkably elegant and well-balanced on the palate with notes of citrus fruits and honeysuckle. Poggio dei Gorleri is also a 'wine resort', complete with an inviting swimming pool and beautifully furnished rooms with views over the vineyards.

www.gorleriwineresort.com; tel 0183 495 207; Via San Leonardo, Frazione Gorleri, Diano Marino; by appointment

04 CANTINE LUNAE

Framed by the Apuan Alps to the north and the rolling hills of Tuscany to the east, the Colli di Luna proves to be a remarkable terrain for producing Vermentino and other varieties. In this easternmost corner of Liguria, Paolo Bosoni has become a legend for the brilliant wines he produces on the 45 hectares (111 acres) of the Cantine Lunae estate. The Vermentino Riserva Etichetta Nera (or Black Label Reserve Vermentino) is a revelation: it exhibits outstanding structure, with aromas of wild herbs, spices and honey. The winery is set among restored 18th-century stone and terracotta-tiled buildings, with an atmospheric wine shop and tasting room, as well as a small museum in the old manor house that pays homage to the people who worked the land in centuries past.

www.cantinelunae.com; tel 0187 693 483; Via Palvotrisia 2, Castelnuovo Magra; Mon–Sat & Sun mornings

05 TERRE BIANCHE

This storied family-run winery dates back to 1870 when Tommaso Rondelli planted the first Rossese vines on the unusual white clay soil above the village of Dolceaqua in western Liguria. Successive generations of Rondelli have worked the land with passion, nurturing vines and continuing a centuries-old winegrowing tradition. Terre Bianche produces several top-notch white wines, including a golden Pigato and a luminous Vermentino. The star of the show, however, is the Rossese di Dolceacqua, a much-lauded ruby-red wine with a rich flavour

04 Francesca Bruna of Bruna winery

05 Pastel hues on the Ligurian coast

06 Antique vintages at Terre Bianche

07 Sestri Levante

profile and notes of cherry, wild berries and spices. It's aged in steel tanks, ensuring that every drop you taste is an unadulterated expression of the terroir. Terre Bianche also rents out holiday apartments nestled among the sloping vineyards set between mountains and sea (which you can explore using the mountain bikes available for guests). *www.terrebianche.com; tel 0184 31426; Località Arcagna, Dolceacqua; daily by appointment*

06 BRUNA

Winemaker Riccardo Bruna bucked the trend with his ambition to make a high-quality Pigato when he began producing wine back in 1970. Until then, this spotted white wine grape (a relative of the better-known Vermentino) was largely an easy-drinking quaff for local consumption. Bruna, however, saw incredible potential in the forest-fringed heights around Ortovero in western Liguria, and within a decade he had won acclaim for his savoury, well-made wines. Today, Riccardo's daughter carries on the tradition, crafting not only brilliantly balanced Pigatos, but also juicy reds made from Granaccia, a grape variety native to Liguria and similar to the Grenache of France's Rhône Valley. *www.brunapigato.it; tel 0183 318 082; Via Umberto I 81, Ranzo; by appointment*

ESSENTIAL INFORMATION

WHERE TO STAY

HOTEL GIANNI FRANZI

In picture-perfect Vernazza, Hotel Gianni Franzi has atmospheric rooms set with antique furniture and simple traditional architecture. Breakfast on the deck delivers sublime sea-drenched views; there's also a small garden under the Doria castle for guest use. *www.giannifranzi.it; tel 0187 812 228; Via San Giovanni Battista 41, Vernazza*

VALLEPONCI

Only 5km (3 miles) from the pretty beach of Finale Ligure, Valleponci feels deliciously wild, tucked away in a rugged Ligurian valley. Horses graze, grapevines bud and the restaurant turns out fresh Ligurian dishes with vegetables and herbs from a kitchen garden. Rooms are simple but show the keen eye of the Milanese escapee owners. *www.valleponci.it; tel 329 315 4169; Località Verse, Val Ponci 22, Finale Ligure*

06

WHERE TO EAT

TRATTORIA DA BILLY

Hidden off a narrow lane in the upper reaches of Manarola, Trattoria da Billy fires up some of the best seafood anywhere in Cinque Terre. Start off with a mixed platter – 12 different hot and cold dishes (octopus salad, lemon-drizzled anchovies, tuna with sweet onion) – then feast on lobster pasta or swordfish with black truffle. *www.trattoriabilly.com; tel 0187 920 628; Via Rollandi 122, Manarola*

CASA E BOTTEGA

Join the day trippers from France at this stylishly bucolic restaurant, cafe, homewares shop and general village epicentre in Dolceacqua. Lunch and dinner dishes are fresh, bold reworkings of local classics: perfect fare for alfresco dining with a bottle of Rossese. *www.ristocasaebottega.it; tel 340 566 5339; Piazza Garibaldi 2, Dolceacqua*

WHAT TO DO

Liguria has hundreds of kilometres of well-signed trails, though some of its most famous tracks lie near Cinque Terre. One memorable two-and-a-half-hour walk heads from seaside Manarola up to the lofty settlement of Volastra (elevation 335m/1099ft), passing through terraced vineyards and olive groves before descending to the clifftop village of Corniglia. You can also arrange guided walks and vineyard visits with Manarola-based outfitter Arbaspàa. *www.parconazionale5terre.it; www.arbaspaa.com*

EVENTS

Held in early April, Sciacchetrail is a challenging 47km-long (29 miles) endurance event that takes to the hills of Cinque Terre. Some 300 runners from around 20 different countries race along dirt paths that skirt past cliff edges and through terraced vineyards, ascending more than 2600 vertical metres (8530ft) over the 10-plus hour race (though winners run it in less than six hours). True to its name, the race pays homage to the famed wine of Cinque Terre, and Sciacchetrà wine tastings, cooking demonstrations and discussions by winemakers are all essential parts of the weekend events. *www.sciacchetrail.com*

07

01

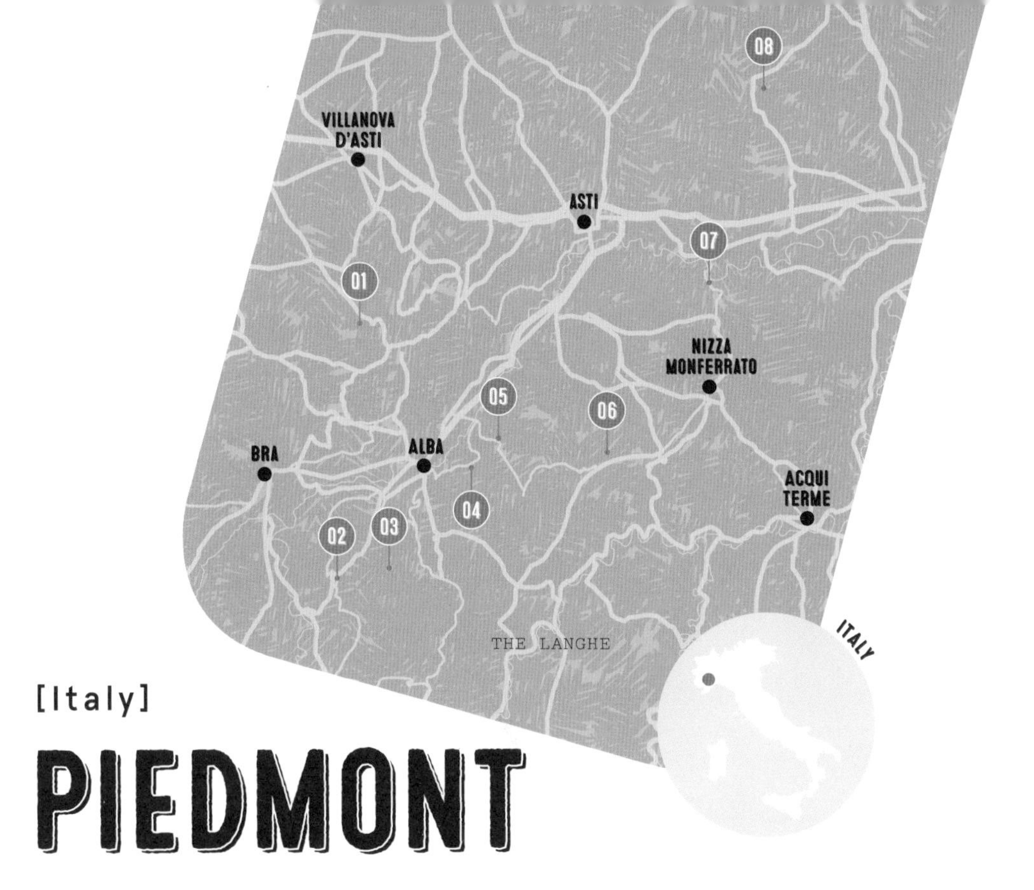

[Italy]

PIEDMONT

Dive into deep, delicious Barolo and Barbaresco red wines in this famously food-loving quarter of northern Italy, before checking out the local truffles.

No region in Italy quite compares to Piedmont's combination of fine wines, gastronomy and beautiful countryside, lying at the foot of the Alps. It would be quite easy to spend a whole trip just wandering through the picture-postcard vineyards and celebrated cantinas of the Langhe, whose hills produce some of the world's greatest, most structured red wines, Barolo and Barbaresco, as well as Nebbiolo, Barbera and Dolcetto. The vineyard landscape is unique and so perfect, that in 2014 it achieved the ultimate honour of being added to the exclusive list of Unesco World Heritage Sites.

But it's worth exploring further to discover the surprising Grignolino and Freisa of the Monferrato region, which are made from native grapes, as well as the delicately sweet Moscato and bubbly Spumante produced around Asti and Canelli.

Meanwhile, a new generation of young *viticoltori* (winegrowers) are bringing to life the rural countryside of the Roero, just across the river from the Langhe, with Arneis and Favorita – both fresh, aromatic whites. Stay in a rustic *agriturismo* where food-lovers can splash out on the ultimate gourmet extravagance: aromatic white truffles, or enjoy simple handmade pasta known as plin, which is stuffed and pinched together.

This is one of the most developed parts of Italy for wine tourism, with numerous winemaker B&Bs and splendid regional *enoteche* (wine shops), where dozens of different wineries are presented in a single tasting. Visitors will quickly realise that the Piedmontesi are reserved people, very proud of their own culture and language. They may not fall over at first to ingratiate themselves with tourists, but you'll soon discover just how hospitable and friendly they are.

GET THERE
Turin-Caselle is the nearest major airport, 73km (45 miles) from Montà. Car hire is available.

01 MICHELE TALIANO

The Tanaro River divides the Barolo and Barbaresco vineyards of the Langhe from Roero, a more biodiverse landscape encompassing farm and woodlands too. Today, a new generation of Roero *viticoltori* are pushing boundaries and producing some exceptional Barbera and Nebbiolo wines. It's when it comes to white wines that the Roero terroir comes into its own, making a serious reputation for the crisp, acidic Arneis and more fragrant Favorita.

Accompany Ezio Taliani on a tour of the vineyard and you embark on an adventure safari on rutted tracks through dense forest before coming out at a breathtaking vista of graphic criss-crossing vines. Make sure you ask him to open a bottle of the intensely aromatic sparkling Birbet, made from Brachetto del Roero, a native grape that is fast disappearing.

www.talianomichele.com; tel 348 734 2873; Corso Manzoni 24, Montà; daily by appointment $

02 CANTINA MASCARELLO BARTOLO

Maria Teresa Mascarello may not have a website to her name, but visitors are certainly made to feel welcome at her tiny cantina in the heart of the medieval wine town of Barolo. The winemakers around here are divided into modernists, who favour single vineyard cuvées, aged in small French *barrique* barrels, and traditionalists who insist on blending different parcels of vines and using huge Slavonian oak casks. Maria Teresa is definitely a traditionalist: she's a fierce defender of Barolo's historic identity, making wines of intense purity and finesse. Working a small 5-hectare (12-acre) estate of prime Nebbiolo vines, she is a far cry from the typical red-faced Piemontese *viticoltore* – a pixie-like figure who looks miniscule as she walks past the towering wooden vats in her cantina.

Tel 0173 56125; Via Roma 15, Barolo; Mon, Tue & Fri by appointment

03 PAOLO MANZONE

Serralunga is a spectacular amphitheatre of vineyards, and Paolo Manzone's *cascina* (farmhouse, cellar

01 Ca' del Baio winery

02 Harvest at Braida estate

03 Barolo wine country

04 Tasting at Braida winery

No region in Italy quite compares to Piedmont's combination of fine wines, gastronomy and beautiful countryside

and *agriturismo*) is hidden away down a zigzag dirt track. A lengthy tasting session with Paolo is the perfect opportunity to understand the complex world of Barolo. He is an innovative *viticoltore*, forever experimenting but never abandoning the traditions surrounding Barolo's unique grape, Nebbiolo. It has been grown here for some seven centuries, and takes its name from the mist that often descends on the vineyards in autumn.

Paolo makes two very different Barolo – the traditional Serralunga, which is aged in large, old oak barrels, and the more modern Meriame, using smaller, new French barrels. He describes his crisp, fresh Dolcetto d'Alba as 'a wine I make for my father – not elegant but rustic, drinkable, like the wine that he sold in demijohns'.

www.barolomeriame.com; tel 0173 613 113; Località Meriame 1, Serralunga d'Alba; daily

04 CA' DEL BAIO

Three generations work together in this idyllic winery nestling in a valley of vineyards. This is classic Barbaresco country, a wine that historically has been the 'little brother' of Barolo, but when you taste this family's vintages, you will discover it can reach equally great heights. The winemaking is in the hands of three dynamic sisters, Paola, Valentina and Federica, who recount 'when our great-grandfather bought the land in 1900, everyone thought he was mad, that it was just worthless woodlands. But he always believed in the potential of the soil and began planting vines.' Don't miss the eminently drinkable Dolcetto – 'great with a pizza,' says Paola with a grin.

www.cadelbaio.com; tel 0173 638 219; Via Ferrere Sottano 33, Treiso; Mon–Sat by appointment

05 Carlo Santopietro of Il Mongetto

06 Twilight over Turin

05 CANTINA DEL GLICINE

Adriana Marzi is a delightfully eccentric woman, but very serious about the award-winning Barbaresco she produces from this small 6-hectare (15-acre) estate. Before the tasting, Adriana takes you through a forbidding blood-red door that leads down to the cantina, what the Piemontese call 'Il Crutin', a natural grotto that is then hollowed out and extended into a maze of damp, cool cellars. This one dates back to 1582, and is like walking into a scene from *Lord of the Rings*, with mushrooms, greedily gobbled up by snails, growing over the damp walls, dark corners stacked with ancient wooden barrels, and alcoves filled with dusty bottles laid down to age. Beware that Adriana always insists visitors try her famous fiery grappa.
www.cantinadelglicine.it; tel 0173 67215; Via Giulio Cesare 1, Neive; Thu–Mon $

06 CANTINE COPPO

The words Asti and Spumante have been famous throughout the world for more than 150 years as the symbol of Italian sparkling wine. The story of Spumante began in Piedmont, near to the town of Asti in medieval Canelli, specifically at the house of Gancia. Sadly the world-renowned Gancia winery is no longer open to the public, following takeover by a Russian vodka company, but there are a host of other spectacular cellars, known as the underground Cathedrals of Canelli, that welcome wine-lovers. Founded in 1892, the family-run Cantine Coppo has one of the most evocative cellars, hewn into Canelli's hills and recognised as a Unesco World Heritage Site. It stores over a million bottles of Spumante, which is made using the same grapes as Champagne – Chardonnay and Pinot Nero – and the historic *metodo classico* process to add bubbles. Be sure to taste its other local speciality, the incredibly fruity Moscato d'Asti.
www.coppo.it; tel 0141 823 146; Via Alba 68, Canelli; daily by appointment $

07 BRAIDA

Braida is forever associated with the name of the late Giacomo Bologna, a mythical figure of Piedmont wine. Planting the then humble grape of Barbera in the unsung region between Asti and Alessandria back in the 1960s, Bologna proved that Piedmont's great wines did not have to be restricted to the Nebbiolo-based Barolo and Barbaresco. Using 100% Barbera and ageing for long periods in small French oak barrels, he produced stunning vintages of what is now Braida's signature full-bodied Bricco dell'Uccelone. Today, this dynamic winery is run by Giacomo's children, Raffaella and Giuseppe, and after a visit to the state-of-the-art cantina, don't miss lunch at the family Trattoria I Bologna.
www.braida.it; tel 0141 644 113; Via Roma 94, Rocchetta Tanaro; Mon–Sat $

08 IL MONGETTO

North of the Langhe, the wilder, under-the-radar region of Monferrato is the place to discover rare indigenous grapes. The brothers Carlo and Roberto Santopietro have converted an 18th-century *palazzetto* (frescoed mansion) into a guesthouse where wines are tasted, and at the weekend local specialities are served in a cosy dining room. They produce surprising reds, such as the fruity but tannic Grignolino, a *vivace* (lively) Cortese, the intense, oak-barrelled Barbera del Monferrato Superiore, and an easy-drinking Malvasia di Casorzo – sweet, fizzy and only 5% alcohol.
www.mongetto.it; tel 347 725 1306; Via Piave 2, Vignale Monferrato; daily by appointment $

ESSENTIAL INFORMATION

WHERE TO STAY

CASA SCAPARONE

A wonderful working farm and vineyard with chic rooms and a raucous osteria serving organic vegan dishes, often accompanied by live music. Kids will love visiting the farm animals. *www.casascaparone.it; tel 0173 33946; Località Scaparone 45, Alba*

CASTELLO DI SINIO

This 800-year-old castle dominates the hamlet of Sinio, surrounded by vineyards producing the finest Barolo wine. You'll find sumptuous rooms and a great welcome by owner Denise Pardini. *www.hotelcastellodisinio.com; tel 0173 263 889; Vicolo del Castello 1, Sinio*

LE CASE DELLA SARACCA

A unique location where six medieval houses have been transformed into a labyrinth of grottoes, suspended glass walkways, swirling metallic staircases and bedrooms with features carved out of the rock. *www.saracca.com; tel 0173 789 222; Via Cavour 5, Monforte d'Alba*

WHERE TO EAT

OSTERIA DA GEMMA

Visit Signora Gemma's osteria to taste her legendary Piemontese *cucina casalinga* (home cooking), where portions of razor-thin *tajarin* pasta are sprinkled with pungent white-truffle shavings. *Tel 0173 794 252; Via Marconi 6, Roddino*

PIAZZA DUOMO

Chef Enrico Crippa has won a coveted three Michelin stars in this futuristic temple of gastronomy in Alba. Sublime cuisine. *www.piazzaduomoalba.it; tel 0173 366 167; Piazza Risorgimento 4, Alba*

WHAT TO DO

Turin was the first capital of modern Italy and is home to ornate baroque palaces, an amazing Egyptian Museum and the Museo Egizio di Torino, as well as a host of historical cafes that have made an art form of the evening aperitivo. *www.museoegizio.it*

CELEBRATIONS

Every weekend during October and November, Alba plays host to its renowned white-truffle festival. *www.fieradeltartufo.org coonawarra.org/event/coonawarra-cabernet-celebrations*

06

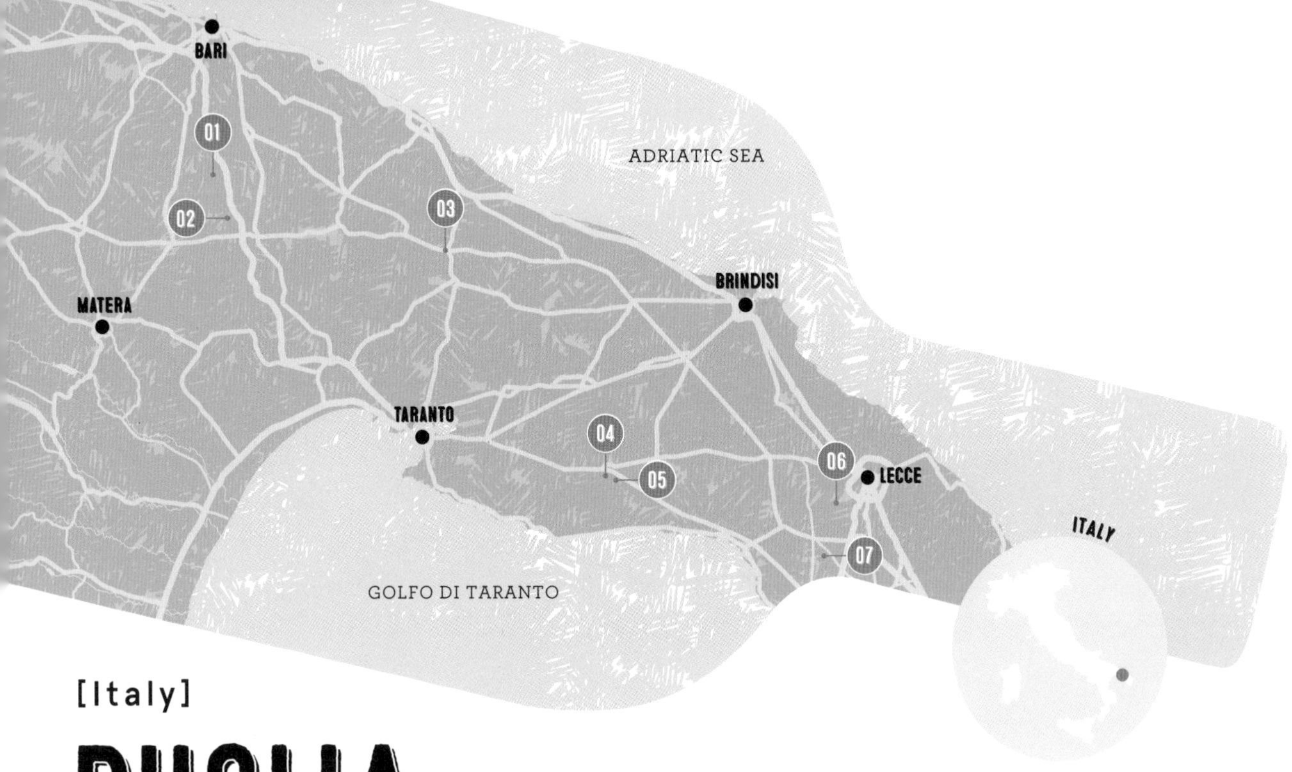

[Italy]

PUGLIA

Follow the sun down through the heel of Italy to olive groves and trulli – and a range of rustic, resurgent wines that accompany the local cuisine perfectly.

Puglia is the largest wine producer in Italy, a quintessentially rural region where cultivating grapes and olives is ingrained in the daily life of the *contadino* (farm worker). But for a long time it has been known for all the wrong reasons, historically supplying wine in bulk to Italy and much of Europe.

Things have changed, though, and today the world has woken up to the wine revolution taking place here. Forget so-called international grapes like Chardonnay, Cabernet and Merlot, and discover unique indigenous grapes – elegant Negroamaro, full-bodied Primitivo, cousin of California's Zinfandel, fruity Minutolo and Malvasia Nera.

The prime vineyards begin just north of Bari stretching down Italy's heel to Brindisi, Taranto, Lecce, Manduria and the Salice Salentino. The climate here is hot but tempered by breezes from the Adriatic and Ionian seas, meaning the wines are intense and strong in alcohol but by no means overpowering. And today the region's small independent estates produce wines of exceptional quality, making use of modern cellar techniques and taking care in the vineyard to limit yield and to cultivate old *albarelli* (bush vines).

Don't expect too much picturesque scenery of hills covered with vines, of the type you'd find in Tuscany or Piedmont. In Puglia, running down through to the southeastern tip of the Italian mainland, landscapes are dominated by miles and miles of flat plains planted with hundreds of thousands of giant, gnarled olive trees, some more than three millennia old. Olive groves and vineyards alike are marked by the unique *trulli* conical white stone huts, some of which are still home to agricultural workers, although others are being converted into seductive B&Bs.

GET THERE
Bari is the nearest major airport, 40km (25 miles) from Acquaviva delle Fonti. Car hire is available.

01 TENUTE CHIAROMONTE

Located just south of Bari, Acquaviva is a medieval town with a long history of winemaking. 'We like to say that this is a mini Rheims,' explains owner and winemaker Nicola Chiaromonte, 'as more than 500 houses have their own cellar where wine has always been made.' Nicola's intense Primitivo Riserva is made from 80-year-old *albarelli*. He is firmly against the trend to lower the alcohol volume of Primitivo to sell to a wider market. 'If you drink a Primitivo that is 13% you will never understand what the wine is about – the grape needs to develop, to mature, and for the Riserva I even favour a partial *appassimento* (desiccating the grape on the vine).'

www.tenutechiaromonte.com; tel 080 768 156; Via Suriani 27, Acquaviva delle Fonti; Mon–Sat $

02 POLVANERA

Polvanera refers to the distinctive deep red soil that surrounds the ancient manor house of innovative *vignaiolo* (winegrower) Filippo Cassano. 'I come from a family of winemakers who for generations produced bulk wine, without the financial means to bottle their own vintages,' Filippo recounts. 'So I am determined to prove that this part of Val d'Itria can make the finest Primitivo as well as great wines from other native Puglia grapes.' He has excavated a quite incredible cellar hewn out of limestone 8m (26ft) underground. This is where he ages his wine for long periods – but there is not a barrel in sight. Filippo bucks the usual trend, refusing to use any wood, preferring to leave the wine in the bottle. The results are spectacular, especially the Polvanera 17, Primitivo in *purezza*, a dizzy 17% alcohol but still fresh and fruity.

www.cantinepolvanera.it; tel 080 758 900; Strada Vic.le Lamie Marchesana 601, Gioia del Colle; Mon–Sat $

03 I PASTINI

While Puglia is making its name right now with Primitivo and Negroamaro red wines, there are also some highly original native white grapes grown in the region.

01 Old town, Bari

02 Tending the vines at I Pastini

03 Harvesting by hand

04 Fermenting grapes at Morella Vini

No estate quite compares to that of the Carparelli family, whose vineyards stretch around the historic city of Locorotondo through Val d'Itria, also known as the Trulli Valley. A tasting includes a visit to the family's 17th-century *masseria* (traditional Puglian farmhouse) and ancient *trulli* still used during the *vendemmia* (harvest).

White wine is their speciality, and the Carparellis are credited with rediscovering the unique indigenous Minutolo grape from the Muscat family, wrongly named for decades as a Fiano from Campania. They vinify and age their wines only in steel vats, also producing Verdeca and the little-known Bianco d'Alessano.

www.ipastini.it; tel 080 431 3309; Strada Cupa Rampone, Martina Franca; Mon–Fri by appointment

04 MORELLA VINI

For the moment, Gaetano Morella and Lisa Gilbee are making genuine garage wines in an industrial warehouse on the drab outskirts of Manduria. But these garage vintages are also winning Italy's top wine awards.

Finding the cantina is a problem, as there is not even a sign, but a tasting is fascinating, as Melbourne-native Lisa explains the experiments she is undertaking: old and new barrels, Nomblot cement eggs and strangely shaped cement vats called 'hippos'. 'We came to Manduria specifically for the ancient *albarelli* vines,' says Lisa. 'So many have already been dug up or abandoned, and we are just trying to save as many as possible because, quite simply, they produce incredible wines.'

www.morellavini.com; tel 099 979 1482; Via per Uggiano 147, Manduria; Mon–Sat by appointment

05 CONSORZIO PRODUTTORI VINI

The role of the cooperative cantina in Puglia's wine history has not exactly been glorious, and many still today prefer the economic security of supplying cheap wine in bulk. A notable exception is the venerable Consorzio Produttori Vini of

Ø5 Bottling time, Apollonio 1870

Ø6 Wine and food pairing at Apollonio 1870

Manduria, the oldest cooperative in Puglia, formed in 1932. Visiting the immense cantina is the perfect introduction to the region's wines. A tasting encompasses not just the iconic Primitivo di Manduria, the historic 'home' of Primitivo, but Negroamaro, white and rosé varietals, and a Fiano Spumante. Beneath the cantina is a labyrinth of cement cisterns, brilliantly transformed into a museum documenting the history of wine and rural life here.

www.cpvini.com; tel 099 973 5332; Via Fabio Massimo 19, Manduria; Mon–Sat

06 APOLLONIO 1870

Marcello and Massimiliano Apollonio run a huge modern winery, the vineyards of which span the prime Negroamaro and Primitivo regions of Salento and Copertino. The family comes from four generations of winemakers who are renowned for ageing Pugliese wines in barrels; and the oenologist of the family, Massimiliano, could well be described as the Wood King.

Touring the cantina, there is not a steel vat in sight, while even the cement cisterns are lined with oak. Massimiliano experiments with different woods for different cuvées – French, American, Slavic, Hungarian and Austrian. He even visits the coopers to order the wood three years before the barrel is made, when it is still a tree. His philosophy? 'Wood is fundamental for me, not just for the *profumo* (scent), but for the colour and stability of the wine, to age to immortality if possible.'

www.apolloniovini.it; tel 0832 327 182; Via San Pietro in Lama 7, Monteroni di Lecce; Mon–Fri $

07 CANTINA SEVERINO GAROFANO

If any master winemaker was responsible for raising Negroamaro to the heights it has reached today, it was Severino Garofano. For 50 years he held the reins at the respected Cupertinum cooperative – still very much worth a visit – moving the conservative *soci* (members) from bulk to quality bottled wine. He also found the time to set up his own winery, a 50-hectare (125-acre) vineyard, which today is in the safe hands of his son, Stefano. This stellar *azienda* (company) is housed in a wonderfully retro 1950s cantina, and the vinification is similarly traditional, still using giant underground cement cisterns. But the wines are absolutely modern, especially the mellow Negroamaro Le Braci, aged for at least seven years, where a short *appassimento* on the vine means alcohol levels rise to almost 15% without affecting the elegance of the vintage.

www.vinigarofano.it; tel 0832 947 512; Località Tenuta Monaci, Copertino; Mon–Fri by appointment $

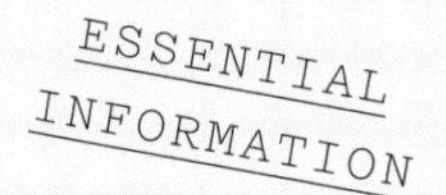

WHERE TO STAY

CANNE BIANCHE
Right at the edge of the beach, this fashionably chic resort has a spa, cooking courses and even offers boat trips to catch *octopi*.
www.cannebianche.com; tel 080 482 9839; Via Appia 32, Torre Canne di Fasano

MASSERIA LE FABRICHE
Five minutes from the sandy beaches of the Ionian Sea, this perfectly restored, 18th-century *masseria* is surrounded by vineyards. Guests stay down below in a secluded olive grove inside modern, minimalist junior suites.
www.lefabriche.it; tel 099 987 1852; C.da Le Fabbriche, SP130, Maruggio

WHERE TO EAT

L'ORECCHIETTA
Favourite haunt of local *vignaioli*, this creative pasta laboratory serves handmade *orecchiette* with *polpette* (meatballs) in rich tomato sauce, plus authentic Pugliese specialities such as *ciceri e tria* (chickpeas and fried pasta).
Tel 0832 705 796; Via Vittorio Veneto 49, Guagnano

OSTERIA DEL POETA
Alberobella is home to Puglia's unique *trulli* stone cottages, a must-see. The perfect place for lunch is chef Leonardo Marco's gourmet osteria.
www.osteriadelpoeta.it; tel 080 432 1917; Via Indipendenza 25, Alberobella

RISTORANTE CIELO
Ostuni looks out over a sea of olive trees, some dating back a thousand years. This elegant restaurant features a gourmet reinterpretation of local cuisine by Michelin-starred chef Andrea Cannalire.
www.lasommita.it; tel 0831 305 925; Via Scipione Petrarolo 7, Ostuni

WHAT TO DO

Don't miss the baroque architecture of Lecce; its labyrinth of lanes forming a maze of ornate churches and grand palazzi.

CELEBRATIONS

Stretching over two months before and after the Lenten carnival period (February/March), the Carnevale di Putignano is both a sacred and profane celebration, combining religious rites with biting political satire.
www.carnevaledi putignano.it

[Italy]

SARDINIA

Discover the secrets of this Mediterranean isle's punchy wines before hitting the fantastic beaches.

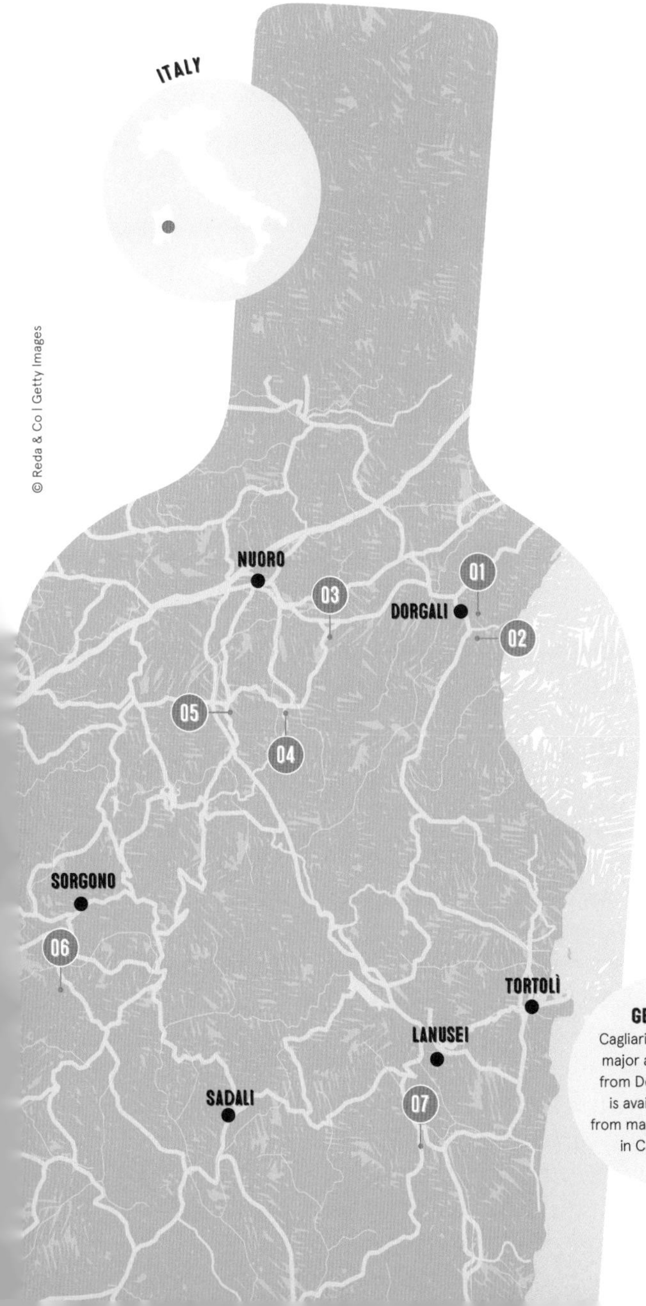

Sardinia is an autonomous region of Italy that sometimes seems to be an independent country all of its own. The stunning sea-and-mountain landscapes contrast Mediterranean beaches with a wild interior of peaks and bare lowlands.

The island produces aromatic Vermentino, Moscato and Vernaccia whites, but it's essentially red-wine country, and that red is Cannonau – explosive, potent and not for the faint-hearted. Known elsewhere as Grenache, Cannonau has grown here for 3200 years, making it the oldest grape in the Mediterranean.

Most of the vineyards, and the best wines, are produced in a triangle of central Sardinia that stretches along an idyllic coastline from Orosei to Barisada, and into the mountainous Barbagia region. Thanks to a centuries-old tradition of pretty much everyone owning small plots of vines and making wine at home, patches of vineyards are scattered around the pastoral landscape. Things changed in the 1950s, when the new *cantina sociale* (cooperative association) grouped several hundred producers; today, new independent winemakers are emerging, creating larger vineyards and making quality Cannonau.

Outside the big cooperative cantinas, wine tourism is in its early days, and it's best to call ahead to arrange estate visits. But once at the cantina, be prepared for a very special welcome from the Sardinian *viticoltori*. Follow their recommendations for a local trattoria and discover a cuisine that is just out of this world: the best cheeses you will taste in Italy, home-cured prosciutto and salami, succulent roast lamb and the unforgettable *porcheddu*, spit-roasted suckling pig – perfect, of course, with an aged Cannonau Riserva.

GET THERE

Cagliari is the nearest major airport, 215km from Dorgali. Car hire is available. Ferries from mainland Italy dock in Cagliari port.

01 On the road in Dorgali

02 Stock up at Gostolai

03 Manicured vines at Poderi Atha Ruja

04 Cala Gonone coastline

01 CANTINA SOCIALE DORGALI

The cooperative winery of Dorgali, just by the seaside resort of Cala Gonone, is the ideal example of how a big *cantina sociale* can put hundreds of *viticoltori* under the same umbrella, producing 1.5 million bottles a year yet succeeding in making an excellent selection of wines.

The unpretentious cantina has something for all wine-lovers, from the holidaymaker who comes in to pick up a bag-in-box of the eminently drinkable *vino da tavola* (table wine), to enthusiasts sitting down for a serious *degustazione* (tasting) of higher-end wines. These include the Cannonau Viniola Riserva and Hortos, a daring blend of indigenous grapes and Syrah, both of which have won Italy's coveted Tre Bicchieri award.

www.cantinadorgali.com; tel 0784 96143; Via Piemonte 11, Dorgali; Mon–Sat $

02 PODERI ATHA RUJA

Pietro Pittalis does not resemble a typical rustic Sardinian *vignaiolo* (winegrower). Immaculately attired, he proudly shows visitors round his tiny vineyard that is as close to perfection as you can imagine. The setting is magnificent, with the lines of manicured vines set against a backdrop of the magical Supramonte mountains that separate these wild Sardinian interiors from the sea.

Pietro is very clear on how to produce a great Cannonau. 'I cultivate a limited number of vines, only 20,000, and each one is cut back throughout the year to grow only five bunches of perfect grapes, which will be made into 20,000 bottles – one for each vine.' In the midst of the vineyards, he has turned a colourful stone cottage into a tasting centre, but this rich, intense Cannonau, aged two years in small oak barrels, really needs another four to five years before reaching maturity.

www.atharuja.com; tel 347 538 7127; Via Emilia 45, Dorgali; daily by appointment $

03 GOSTOLAI

Oliena is the unofficial capital of Cannonau country, but you need to head into the industrial outskirts to discover the modern cantina of Tonino Arcadu. If Tonino seems more interested in talking history and poetry, it is no surprise to learn he was a school teacher before devoting himself to his vines. 'Lots of people in Sardinia want to drink young wines,' he says. 'Well, I don't. I like to age my wines to see how they develop and I am not in a hurry.' The vintages are certainly out of the ordinary, especially a sensational 2012 Riserva D'Annunzio, named after the legendary Italian adventurer, Gabriele d'Annunzio, who visited Oliena as a 19-year-old reporter and immediately became a lifelong aficionado of Cannonau.

www.gostolai.net; tel 0784 285 374; Zona P.I.P, Oliena; Mon-Fri & by appointment Sat-Sun $

04 CANTINE DI ORGOSOLO

The wild mountain town of Orgosolo is a symbol of the inherent independence that Sardinian people claim as their right, marked by the *murales*, over 200 huge political frescoes covering the walls of most of the houses.

Orgosolo is also home to a genuine garage winery producing some of the most original Cannonau on the island. Eleven years ago, 17 diverse smallholders – including a tobacconist, an electrician, a hospital worker and a shepherd – grouped together to make an artisan Cannonau. Although they get occasional advice from a top oenologist, this merry band prefer to spend hours sitting around steel vats and oak barrels in their recently constructed cantina, fiercely discussing vinification and ageing.

'We make one wine that has stayed in the barrel for only three months, easy to drink straight away, and a Riserva that ages for three years,' says one *viticoltore*. 'And if we don't sell them, then we'll just drink them ourselves!'

www.cantinediorgosolo.it; tel 333 380 5605; Via E. Mattana, Orgosolo; Mon-Fri & by appointment Sat-Sun $

05 Tonino Arcadu of Gostolai

06 Sardinian vineyards

05 AZIENDA GIUSEPPE SEDILESU

Giuseppe Sedilesu bought a small plot of Cannonau vines 40 years ago, which produce a stunning range of Cannonau that today are among the award-winning stars of the Sardinian wine scene. The family is moving with the times by producing organic and biodynamic wines, using oxen to plough the soil.

The family oenologist, Francesco Sedilesu, says, 'We concentrate almost exclusively on red wines, which are intense and complex.' The only surprise is that during the tasting, instead of the usual cheese and salami, there is a plate of bitter chocolate, a perfect complement to these powerful vintages.

www.giuseppesedilesu.com; tel 0784 56791; Via Vittorio Emanuele II 64, Mamoiada; Mon–Fri

06 LE VIGNE DI FULGHESU

While Cannonau is not that well known outside Sardinia, the obscure Mandrolisai denomination ranks as an even rarer discovery. Winemakers around the villages of Atzara and Meana have the opportunity to make Mandrolisai, a unique blend of Cannonau and two local grape varieties, Muristellu and Monica.

Stop off in Atzara at the *enoteca* (wine shop) of the excellent Fradiles winery, then trek out to the cantina of Peppe Fulghesu who makes unforgettable wines, many of them natural, producing an incredible explosion of fruit. His vineyard lies at the end of a rutted track, but the reward is a terrific welcome and a tasting accompanied by his home-cured salami and freshly made ricotta cheese. And it doesn't need much to encourage Peppe to take visitors for a hike up the hill to see a rare *nuraghe*, one of Sardinia's famed prehistoric stone huts dating back to 1000 BC.

Tel 0784 64320; Via Su Frigili Cerebinu, Meana Sardo; daily by appointment

07 ANTICHI PODERI DI JERZU

The Jerzu region, located between the sandy beaches of the Bay of Ogliastra and the Gennargentu mountains, has a long tradition of wine production, and has the right to produce its own Cannonau denomination.

Founded in 1950, this dynamic *cantina sociale*, with its 430 soci, is very active in Jerzu's social life, especially each August when it organises the Sagra del Vino, a week-long music and wine festival.

The huge winery dominates the town and is marked by a tower that has been converted into a tasting room with panoramic views. And there is a lot to taste here because, as is the case with each *cantina sociale*, there is plenty on display – including Sardinian speciality, Mirto, a lethal liqueur made from myrtle berries.

www.jerzuantichipoderi.it; tel 0782 70028; Via Umberto 1, Jerzu; Mon–Sat

ESSENTIAL INFORMATION

WHERE TO STAY

SU GOLOGONE

Though it's lost in the middle of the countryside, this luxury resort has a restaurant, a pool and spa and a museum-standard collection of traditional arts and crafts. *www.sugologone.it; tel 0784 287 512; Località Su Gologone, Oliena*

AGRITURISMO CANALES

This rustic mountain hideaway overlooks an emerald-green lake. Fresh cheeses are delivered each morning, and wine comes from its own vineyard. *www.canales.it; tel 0784 96760; Località Canales, Dorgali*

DOMUS DE JANAS SUL MARE

A simple family hotel with a stunning location by a bay with an ancient watchtower that protected the village from marauding pirates in medieval times. *www.domusdejanas.com; tel 0782 28081; Via della Torre 24, Bari Sardo*

WHERE TO EAT

RISTORANTINO MASILOGHI

A favourite hangout of Oliena winemakers, Gianfranco Maccareno's romantic trattoria features the best *porcheddu* that you will ever taste. *www.masiloghi.it; tel 0784 285 696; Via Galiani 68, Oliena*

GICAPPA

This lively, casual restaurant serves simple cuisine, making use of more unusual local produce such as mountain ferns, wild boar and, for fans of nose-to-tail eating, lamb sweetbreads, brain and *coratella* (lungs, heart and liver). *www.gicappa.it; tel 0784 288 024; Corso Martin Luther King 4, Oliena*

WHAT TO DO

From the Cala Gonone seaside resort, hire a boat to sail into the mysterious Bue Marino cave which is filled with stalactites and stalagmites. Afterwards soak up the sun on on idyllic Cala Luna beach.

CELEBRATIONS

The hamlet of Mamoiada hosts an extraordinary animist carnival just before Lent. Villagers are transformed into wild *mamuthones*, fearful-looking creatures clad in masks and thick black sheepskin fleeces draped in heavy cowbells. Bonfires are lit and a large quantity of wine is drunk.

01

[Italy]

SICILY

Ancient grape varieties, adventurous winemakers and a myth-shrouded mountain combine to create one of Europe's most exciting wine scenes: Etna.

So, what does a volcano taste like? And who would make wine on the slopes of Europe's tallest active volcano? Those are the kind of questions you'll find yourself asking on this tour around Sicily's Mt Etna.

Shipwrecks show that Sicily has been exporting wine since at least 2 BC, first in clay amphorae then wooden casks. The Greeks then the Romans made wine on this crossroads of the Mediterranean. Now a new generation of winemakers has revitalised the vineyards around Mt Etna. They've been attracted by a unique combination of factors: interesting indigenous grape varieties, a well-drained soil rich in minerals coupled with a high-altitude climate, and a winemaking culture based around *contrada* (parcels of land) and *palmenti* (old stone cellars) found only on Sicily.

This trail follows an ancient Roman road (now the SS120) as it curves around the north side of Mt Etna, passing, (west to east) through the wine towns of Randazzo, Passopisciaro, Solicchiata and Linguaglossa, before bearing south to finish in the region around Milo that specialises in an unforgettable white wine. The major town in this corner of the island is Taormina, famed for its Greek amphitheatre. But it too is upstaged by the view of Etna: red earth, black rock and a smouldering white peak.

The 3326m volcano is an ever-present companion on this trail. Lava flows from past eruptions leave walls of blackened rock and pumice in many of the vineyards. High above the searing heat of the Sicilian summer, the altitude of Etna's Denominazione di Origine Controllata means that the grapes ripen slowly as they grow. They share the land with olive trees and orchards of almond, hazelnut, walnut, fig, peach and citrus trees. Fresh seafood is landed on the coast. And the food and wines combine to create a local cuisine that tastes out of this world.

GET THERE
The most convenient international airport is Catania, an hour's drive south. Trains from mainland Italy reach Palermo to the west.

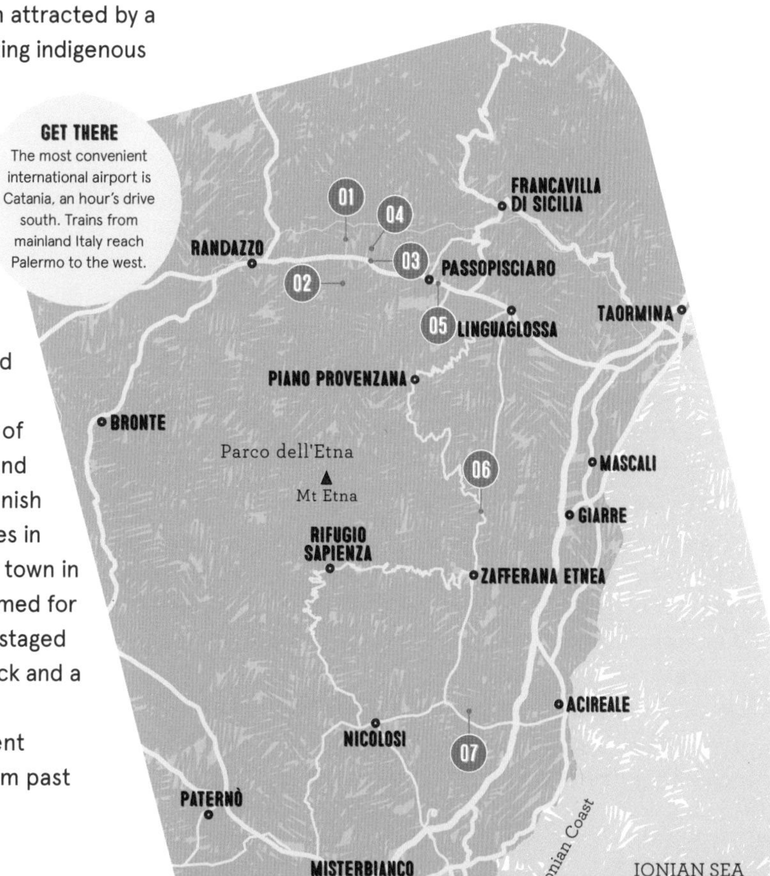

01 FATTORIE ROMEO DEL CASTELLO

For a first impression of how close the relationship between the volcano and the vineyards can be on this north slope of Etna, start at Chiara Vigo's organic farm on the outskirts of Randazzo. The vineyard is part of her family's centuries-old Romeo del Castello estate and in 1981 a lava flow destroyed 21 hectares (52 acres) before changing course away from the historic house, leaving a hinterland of blackened rock. After studying at the University of Bologna – she was tutored by Umberto Eco and published a thesis on wine labels – Chiara returned home in 2007 to take responsibility for the family's heritage: 'Wine was always made here and we now produce three red wines, all from old-growth Nerello Mascalese vines.' Chiara's parcel of land is named Contrada Allegracore (meaning 'happy heart') and she believes in letting the grapes speak for themselves with minimal intervention. 'The soils around Etna come from different eruptions so each parcel produces different wine – that's why Etna wine producers specify the parcel on the bottle.'

www.romeodelcastello.it; tel 095 799 1992; Contrada Allegracore, Randazzo; by appointment

02 PLANETA

If Fattorie Romeo del Castello epitomises small-scale Sicilian production, Planeta is an example of big investment done well. The Italian wine group purchased plots of land in 2008 near Passopisciaro, the epicentre of Etna's wine country. Under the guidance of long-standing head winemaker Patricia Tóth, Planeta has restored old buildings, planted Nerello Mascalese and Carricante vines, nurtured long-lost varieties – the *reliquie* or relics – and developed vineyards such as Sciaranuova. Some terraces were turned into an open-air theatre that is the venue for the annual Sciaranuova arts festival. The new winery and tasting space, sympathetically constructed from lava rock, opened in 2012 in the *contrada* of Feudo di Mezzo. Tasting

01 Mt Etna and Taormina

02 Dive in at Barone di Villagrande

03 Planeta's Feudo di Mezzo winery

04 Wine tasting at Planeta

experiences include a vertical tasting of three wines at the Sciaranuova *palmento*, plus a visit to the lava flows.
www.planeta.it; +39 0925 1955465; winetour@planeta.it; Contrada Sciaranuova, Passopisciaro; by appointment

03 FRANK CORNELISSEN

'There are many reasons to be attracted to a place,' says Frank Cornelissen. 'The northern valley of Etna has something mystical: the obvious natural beauty but also its energy and contradictions, snowed under in the winter, hot and dry during the summer. We're living in the southern Mediterranean but on the side of a mountain. The result is the incredible elegance and structure of the wines.' Sometimes it takes an outsider to appreciate a place fully: Frank Cornelissen is regarded as a radical but the Belgian winemaker, who is dedicated to creating the most natural wines possible, is responsible for many of Etna's most interesting wines. He bottles nine of his wines according to the vineyards or *contrada* in which the grapes grew. The Munjebel red from Monte Colla, for example, comes from Nerello Mascelese vines planted in 1946 on a steeply terraced site right before Mt Etna. All Frank's wines are produced without chemical preservatives and using indigenous yeasts. He's very happy to explain his processes and philosophy at his Passopisciaro cellar, after a tour of a vineyard.
www.frankcornelissen.it; info@frankcornelissen.it; Via Canonico Zumbo 1, Passpisciaro; by appointment

04 VINI FRANCHETTI - PASSOPISCIARO

Adventure runs in founder Andrea Franchetti's veins; Ernest Hemingway was a friend of the famed Tuscan Italian family. After Andrea left school he cycled and hitched his way to Afghanistan before returning to Italy by way of Manhattan. In 2000, having married his Sicilian girlfriend, and drawn by the drama of Etna's setting, he began to restore an old farm and cellars on its

05 Pietradolce's modern winery

06 Harvest in the Pietradolce vineyards

slopes just above the wine town of Passopisciaro. He works mainly with the local Nerello Mascalese variety, which grows at up to 1000m above sea level so the grapes are late to ripen - temperatures here are up to 15°C (59°F) cooler than on the coast and downright chilly at night during September and October before harvest. Tours take place twice daily during weekdays and include a vineyard visit.
www.vinifranchetti.com; 094 239 5449; Contrada Guardiola, Castiglione di Sicilia; by appointment

05 PIETRADOLCE

Look around the smart new winery and tasting room at Pietradolce and you get the sense that the Faro family have long been Etna pioneers. 'My grandfather was a small-scale winemaker,' says owner Michele Faro. 'But we were convinced the territory had enormous potential - although it was not certain that you were making the right choice!' Pietradolce's ambition is to produce elegant wines representative of the land - look for gentle tannins in the Nerello Mascalese with red cherry and raspberry fruit, quite like a pinot noir from Burgundy. Old, pre-phylloxera vines are cultivated in wild, terraced vineyards in the local 'bush' or alberello method, surrounded by butterflies and flowers. 'The grapes are fermented, for the most part, in raw cement vats that recall the old traditional fermentation vats but with modern methods of temperature control. So tradition and innovation come together. We are really happy with the result.'
www.pietradolce.it; 344 064 0839; Contrada Rampante, Solicchiata; by appointment

06 BARONE DI VILLAGRANDE

As you travel around Etna clockwise, Sicily's signature white wine, made from the local Carricante grape, comes to the fore. Etna's east side, notably cooler and cloudier, is where Carricante flourishes. For 10 generations the Nicolosi family - Sicilian nobility since the King of Naples made Don Carmelo a baron - has grown grapes in the vineyards of the Bosco Etnea region among their oak and chestnut forests. The winery is well established for visits and a guided tasting gives the lowdown on this unique wine: it's a savoury, almost saline, experience, with aromas of white fruits, almonds and a fresh minerality that makes it a perfect match for seafood and cheese; low alcohol levels also make it an especially sociable wine (a good comparison would be the Albariño wines from Rias Baixas in Spain (p249).
www.villagrande.it; 095 708 2175; Via del Bosco 25, Milo; Mon-Sun 11am-3.30pm, 6.30pm-10pm; reservations required email winetour@villagrande.it

07 BENANTI

Finish your tour of Etna at Benanti, one of the original new-wave wineries that inspired local gurus like Salvo Foti and Alberto Graci. The estate was revitalised in 1988 when Guiseppe Benanti decided to take winemaking seriously again, after the vineyards had languished for decades. It's based in Milo, the heart of Carricante country, and is now run by twin sons Salvino and Antonio. Benanti's benchmark white wine is the zesty Pietra Marina, made from Carricante and aged for three years (and will age well). Look for mineral lemon fruit in this Sicilian classic.
www.vinicolabenanti.it; 095 789 0928; Via Guiseppe Garibaldi, 361, Viagrande; by appointment

ESSENTIAL INFORMATION

WHERE TO STAY

HOTEL FEUDO VAGLIASINDI

Converted from a large *palmento* just outside Randazzo, this art nouveau manor house is drenched in wine history, with original barrels from the 19th century still lined up in the cool cellars. Olive trees and Nerello Mascalese vines are cultivated around the manor and fresh, local produce, including the hotel's own olive oil, is used in the kitchen. Lots of volcano-oriented activities, including bike and horseback rides, canyoning and 4WD tours are available. *www.feudovagliasindi.it; tel 095 799 1823; Contrada Feudo Sant'Anastasia, Strada Provinciale 89, Randazzo*

IL NIDO DELL'ETNA

The town of Linguaglossa makes a good base for wine-touring around Etna's north side (look also at Passopisciaro, and many of the wineries offer rooms too). This clean, contemporary three-star hotel is within walking distance of the town centre's restaurants and has views of Etna. *www.ilnidodelletna.it; tel 095 643404; Via Matteotti, Linguaglossa*

WHERE TO EAT

CAVE OX

Entering Cave Ox in Solicchiata is like finding a back door to Etna's wine scene, particularly the natural wine movement. Owner Sandro Dibella is a friend of Frank Cornelissen and they share an interest in traditional Etna winemaking methods. Lots of other local winemakers are also regular visitors, bringing along their latest releases to share. Pizza is served indoors or outside in a garden and is accompanied by an epic wine list. The place is closed on Tuesdays. *www.caveox.it; tel 0942 986 171; Via Nazionale, 159, Solicchiata*

THINGS TO DO

FERROVIA CIRCUMETNEA

This narrow-gauge railway, dating from 1888, runs around the perimeter of Etna, with a section forming part of Catania's Metro. If you depart Randazzo and head clockwise towards Riposto you'll pass through Etna's prime vineyards, lava fields and lemon groves. You can hop on and off at wine villages (six departures daily, Mon-Sat).

ETNA WINE LAB

Freestyle Sicilian driving may be off-putting: a guided tour can allow you to focus on the wine and let someone else deal with the driving. Etna Wine Lab offers a range of experiences taking in Etna's food and wine, meeting producers along the way. *www.etnawinelab.it*

CELEBRATION

CONTRADE DELL'ETNA

A part-tasting, part-party that was first held in 2008 at the instigation of Andrea Franchetti and a couple of dozen local producers and friends. Now the guest list spans a couple of thousand people and includes sommeliers and wine fans from around the world, drawn by the opportunity to taste wine from 100 Etna producers - and to have a good time.

06

01

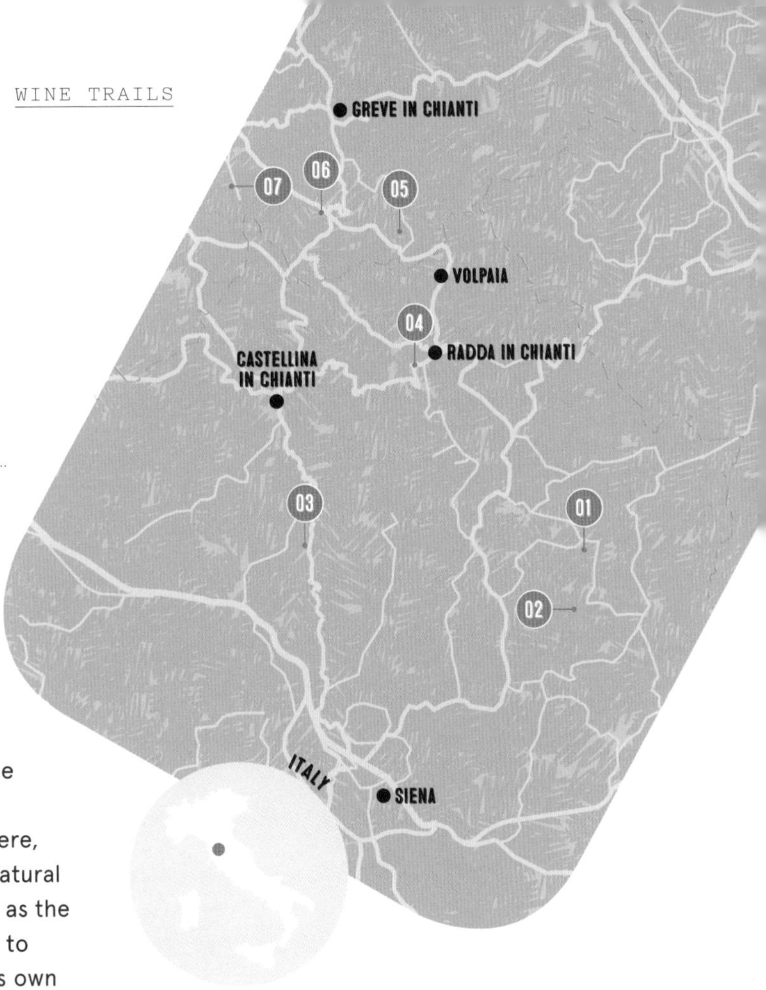

[Italy]

TUSCANY

Due to its swaying cypresses, hilltop villages and exceptional red wines, Tuscany exudes an old-fashioned romance that's hard to resist.

Wine and Tuscany are so closely associated that, for many, the vintages produced in the Tuscan hills symbolise all the glamour and style of Italy in the same way that Champagne evokes France.

It was the Etruscans who first made wine here, using huge terracotta amphorae that some natural winemakers are rediscovering today. As early as the 5th century BC, Tuscan wines were exported to France and Greece, and Florence founded its own Wine Merchants Guild in 1282. So Tuscany has always been the ambassador of Italian wine, from ancient times through to the days when a straw-covered flask of Chianti featured as house wine across the globe. And today, Tuscany's wine has moved onto the wine lists of the world's finest restaurants, with the top Tuscan producers appearing alongside those of Bordeaux and Burgundy.

Tuscany is the perfect location for producing outstanding red wine with the native Sangiovese grape, but centuries of tradition came under threat as winemakers started blending it with 'international' varietals like Cabernet Sauvignon and Merlot. These are the so-called Super Tuscans, wines produced outside the ancient regulations that offer immediate, accessible quality that is easier to sell internationally. While some Super Tuscans, primarily in the Maremma, have made their mark and are here to stay, now the mood is returning to traditional winemaking based purely on the potential of Sangiovese. That is definitely the case in the historic Chianti Classico region, where the story of Tuscan winemaking began.

With vineyards nestling in undulating hillsides and valleys from the outskirts of Florence all the way to the Mediterranean, it is difficult to know quite where to start when planning a Tuscan wine journey. Exclusive wines such as Sassicaia and Ornellaia are produced in the relatively new maritime vineyards of windswept Maremma.

The medieval cities of Montalcino and Montepulciano continue to impose strict rules on the making and ageing of their venerable Brunello and Vino Nobile vintages. But the Chianti Classico region remains the beating heart of Tuscany.

GET THERE
Pisa is the nearest major airport, 120km (75 miles) from Gaiole di Chianti. Car hire is available.

GREVE IN CHIANTI
07
06
05
VOLPAIA
04
RADDA IN CHIANTI
CASTELLINA IN CHIANTI
03
01
02
ITALY
SIENA

01 CASTELLO DI BROLIO

The story of Chianti begins at the enchanting Castello di Brolio. It may have become something of a Disneyland castle, but it remains a must-see stop-off.

While wine in Tuscany dates back to Etruscan times, it was Barone Bettino Ricasoli, former prime minister of Italy and enthusiastic winemaker, who is credited with creating the blend of grapes that produce the distinctive personality of Chianti Classico, back in 1872. This perfect expression and interpretation of the Sangiovese grape has survived until today.

The Ricasoli dynasty owned Brolio from 1141, along with half the countryside between Florence and Siena, and the present descendant Francesco, the 32nd Barone, has restored Brolio's reputation as a respected winemaker.

www.ricasoli.it; tel 0577 730 220; Località Madonna a Brolio, Gaiole in Chianti; daily

02 LE BONCIE

Giovanna Morganti, along with cult French winemaker, Nicolas Joly, was one of the founders of Vini Veri, which today has grown into the influential 'natural wine' movement. She is an uncompromising *viticoltrice* (winegrower), planting her vineyard from scratch in 1990. The freestanding *alberelli* (bush vines) of her tiny vineyard resemble immaculate bonsai trees surrounded by a jungle of wild plants and weeds. Her work in the cantina is exceptional, fermenting in open-topped tanks with regular *batonnage* (stirring of the lees).

Giovanna makes just one wine, Le Trame, a *vino da tavola* that she has taken deliberately outside the official DOCG (*Denominazione di Origine Controllata e Garantita*) appellation, and is unlike any other Chianti Classico you will taste – incredibly intense, complex and concentrated. While 90% of the grapes are Sangiovese, she adds little-known local varietals such as Mammolo, Colorino and Fogliatondo, and is fiercely critical of fellow winemakers who have been influenced to add in so-called 'international grapes' Merlot and Cabernet Sauvignon.

Ø1 Villa Pomona wine tasting

Ø2 Siena skyline

Ø3 Chianti country

Ø4 Winery in Chianti

Ø5 Monica Raspi in the cantina at Villa Pomona

Tel 0577 359 383; Località San Felice, Castelnuovo Berardenga; daily by appointment

03 VILLA POMONA

Traditions die hard in Chianti country, as visitors will quickly understand when they meet the passionate, down-to-earth winemaker Monica Raspi in the charming *fattoria* where she makes wine and olive oil. Guests stay in an ancient olive mill converted into holiday apartments.

Monica explains, 'I was born here, but went off to Florence to study as a veterinarian. But as soon as my mamma said she was going to sell the villa and our vineyards I couldn't bear to lose our family heritage.' She abandoned her career, did a crash course in oenology, started making her own wines in 2007, 'and mamma looks after the bed and breakfast.' While her tiny cantina is a cluttered mix of giant wooden *botte* (barrels), steel and cement vats, she has recently obtained official organic certification for the vineyard.

www.facebook.com/fattoriapomona; tel 0577 740 473; Località Pomona 39, Castellina in Chianti; daily by appointment $

04 VAL DELLE CORTI

A wine tasting with Roberto Bianchi often ends up with an impassioned discussion over several bottles of his outstanding wines accompanied by a plate of delicious Tuscan sausage and cheeses. This feisty artisan *vignaiolo* (winemaker) makes a supple *vino da tavola*, perfect for drinking young, a tannic but elegant Chianti Classico bursting with fruity flavour, and a Riserva only when he feels the harvest merits it. His organic vineyard is only 6 hectares (15 acres), and in his chaotic garage cantina are only aged barrels, because 'there is already enough tannin in the Sangiovese grape.'

www.valdellecorti.it; tel 0577 738 215; Val delle Corti, Località La Croce 141, Radda in Chianti; Mon-Sat by appointment $

05 FATTORIA DI LAMOLE

The road to Lamole weaves through thick forest before emerging at one of the most beautiful villages in the Chianti region, some 600m (2000ft) above sea level.

06 Cosimo Gericke of Rignana winery

07 Fattoria Rignana

08 Piccolomini Library, Siena Cathedral

Paolo Socci is a fiercely traditionalist winemaker and in the cantina he favours giant old wooden barrels: 'I want my wine tasting of Sangiovese tannin, not oak.' His 2013 Gran Selezione Vigna Grospoli is sensational, made solely with grapes grown in the micro-vineyard within the 7km (4 miles) of centuries-old *terrazzi* that Paolo has painstakingly rebuilt. He has also renovated a small medieval hamlet of cottages at the edge of Lamole into a comfy B&B for visitors.

www.fattoriadilamole.it; tel 055 854 7065; Lamole, Greve in Chianti; daily by appointment $

06 FONTODI

Down below Panzano lies a showpiece vineyard planted in a sunny amphitheatre known as La Conca d'Oro. It is the heart of the Fontodi estate that covers 90 hectares (220 acres) of vines and 30 hectares (75 acres) of olive groves. A 50-strong herd of Tuscany's iconic Chianina cows completes this far-sighted and sustainably run organic farm.

Unlike many of Chianti's big wineries, a visit here is a casual affair. The Manetti family bought Fontodi over 50 years ago and are experimenting with ageing a wine in an *orcio* (terracotta amphora) jar, a technique used by the Greeks and Romans. Their modern flagship wine, Flaccianello della Pieve, is 100% Sangiovese, while the luscious Vin Santo is made using grapes that have been straw-dried for five months and then barrel-aged for six years.

www.fontodi.com; tel 055 852 005; Panzano in Chianti; Mon-Sat by appointment

07 RIGNANA

Head off into the unknown along the *strada bianca*, the notorious dirt tracks criss-crossing the Tuscan countryside, and you'll discover romantic villas, vineyards, farms and restaurants. One of these hidden jewels is Rignana, the magical winery of the dashing Cosimo Gericke. His 18th-century frescoed villa has sumptuous guest rooms, an olive mill converted into a trattoria, a pool at the edge of an olive grove and even a medieval chapel used for weddings.

Cosimo produces an absorbing mix of organic Chianti Classico, a tannic Merlot aged in small oak barrels, a rare white wine using Sangiovese, and a fruity, light Rosato, most of which gets consumed in sunset aperitifs at the villa itself.

www.rignana.it; tel 055 852 065; Via di Rignana 15, Greve in Chianti; daily by appointment

ESSENTIAL INFORMATION

WHERE TO STAY

FATTORIA LA LOGGIA

This sprawling medieval wine and olive-oil *fattoria* has been transformed into an idyllic bolthole with spacious rooms, a heavenly pool and sculptures and paintings everywhere thanks to the artists-in-residence programme. *www.fattorialaloggia.com; tel 055 824 4288; Via Collina 24, San Casciano in Val di Pesa*

LE MICCINE

Canadian *vigneronne* Paula Papine Cook settled in Chianti a few years ago, and her wines are already winning top awards. The guest villa next to her cantina features an inviting infinity pool. *www.lemiccine.com; tel 057 774 9526; Località Le Miccine 44, Gaiole in Chianti*

WHERE TO EAT

ANTICA MACELLERIA CECCHINI

Dario Cecchini is the unofficial King of Chianti, a dramatic master butcher famed for his *bistecca alla fiorentina* – T-bone steak. A cheap and cheerful trattoria, Dario Doc, is at the back of the butchers. *www.dariocecchini.com; tel 055 852 020; Via XX Luglio 11, Panzano in Chianti*

A CASA MIA

This village osteria serves huge portions of *cucina casalinga* (home cooking) such as tagliatelle with porcini, presented by the rock'n'roll hosts, Cosimo and Maurizio. *www.acasamia.eu; tel 055 824 4392; Via Santa Maria a Macerata 4, Montefiridolfi*

BAR UCCI

Volpaia is a dreamy medieval village high up in the hills of Chianti. Tuck into Tuscan specialities on Ucci's sunny terrace. *www.bar-ucci.it; tel 0577 738 042; Piazza della Torre 9, Volpaia*

WHAT TO DO

Take a break from the estate visits and wine tastings and spend a relaxed day exploring the Renaissance palaces, churches and museums of the nearby town of Siena.

CELEBRATIONS

In September, make your way to either Panzano or Greve in Chianti, which both host week-long wine festivals.

08

[Italy]

VENETO

Its most famous export is Prosecco, party drink par excellence, but Veneto is also home to a multitude of quaffable reds, from light and fresh to hefty hitters.

Prosecco, made from the Glera grape, seemed to become the world's favourite bubbly almost overnight. The picture-perfect vineyards that produce it stretch from the border with Friuli to Valdobbiadene. They're a paradise for wine travellers, with friendly winemaker B&Bs and rustic local osterias.

The Veneto has made its mark on Italy's wine map as one of the biggest wine-producing regions, featuring a diverse, high-quality range of both reds and whites. From the volcanic hills around Padua, the Colli Euganei, come a variety of wines, from elegant barrel-aged Merlot and Cabernet Franc to the surprisingly fizzy Fior d'Arancio, whose unique orange-blossom aroma rivals a Moscato from Asti. Meanwhile the crisp, white Soave is hitting new levels of excellence.

Then from Verona to Lake Garda you enter Valpolicella country, where wine produce has shifted from the kind you would order in a pizzeria to some of Italy's finest red wines. Based around Corvina, a native grape, Valpolicella caters for every taste, from the Classico, young, fresh and easy to drink, to the Ripasso, where grape pomace has been macerated to add body. And then comes Amarone. This is quite simply a unique wine; the grapes are dried for three to four months – the *passito* process – before fermentation begins, then aged mainly in oak for a minimum of three years before going on sale. Tasting it is a serious business, as alcohol content can rise to a heady 17%. It's the perfect complement to traditional Veneto dishes, such as slow-cooked wild boar stew or, even better, used as an ingredient in a memorable *risotto all'Amarone*.

GET THERE
Venice Marco Polo is the nearest major airport, 64km (40 miles) from Premaor. Car hire is available.

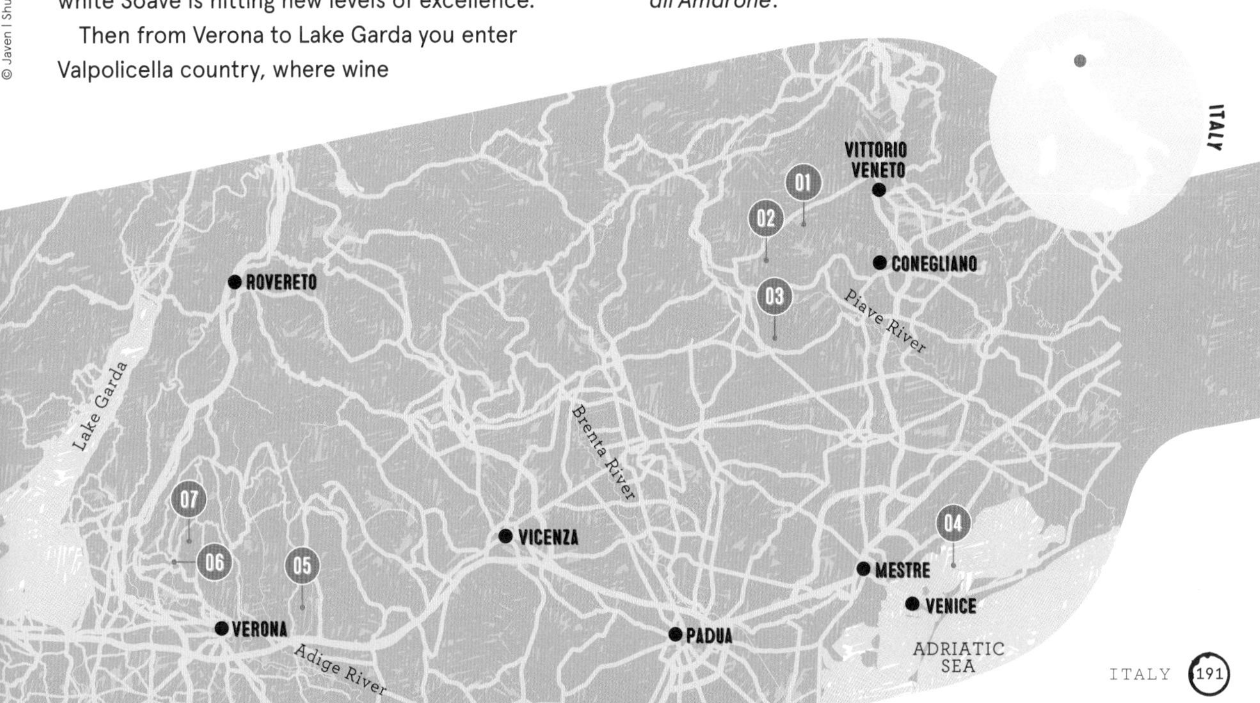

01 GREGOLETTO

The Prosecco vineyards that run along verdant valleys between Conegliano and Valdobbiadene are some of the most beautiful in Italy, and in the sleepy village of Premaor, the Gregoletto family have been cultivating vines since 1600.

A tasting here spans not just an excellent Brut Prosecco and the signature Frizzante, but also white wines from little-known local grapes like Verdisio and Manzone Bianco. You are bound to get caught up in conversation with Giovanni Gregoletto, a wonderfully eccentric character, as happy discussing poetry, music and philosophy as wine. His lunchtime canteen is Trattoria al Castelletto, where slick executives from the nearby Benetton offices rub shoulders with Prosecco *viticoltori* (winegrowers).

www.gregoletto.it; tel 0438 970 463; Via San Martino 81, Premaor; daily by appointment $

02 CASE COSTE PIANE

While the Prosecco region is enormous today, selling even more bottles than Champagne, the jewel in the crown of this bubbly is a minuscule terroir just outside Valdobbiadene. Winemakers battle to own a small parcel of the 107 hectares (265 acres) that constitute the elegant Cartizze, Prosecco's own Grand Cru and one of the wine world's most expensive pieces of real estate, coming in at around €1,000,000 per hectare (2.5 acres).

But the village also has an artisan winemaker equally famous in his own way. Loris only makes 70,000 bottles a year of his wonderful Casa Coste Piane, but you'll find it stocked in Italy's top restaurants. This rustic Prosecco is naturally fermented in the bottle – *sur lie* or *col fondo* – meaning it is *frizzante* (crisp) rather than bubbly, slightly cloudy and best served decanted.

Tel 0423 900 219; Via Coste Piane 2, Santo Stefano di Valdobbiadene; Sat by appointment

03 INAMA

The return of Soave to the family of grand Italian white wines after decades of overproduction has been achieved largely through the efforts of historic cantinas

01 Verona

02 Grape picking at Venissa

03 Venissa's micro-vineyard on Mazzorbo island

04 Farmhouse at Valentina Cubi

including Pieropan, Gini and, perhaps above all, Inama.

Giuseppe Inama founded his estate in the 1960s, and the family's award-winning wines were created first by his son, Stefano. Nowadays, grandsons Matteo, Alessio and Luca have joined the family business too. The Soave region's volcanic vineyards, planted on basaltic lava since Roman times, are perfect for the local Garganega grape.

A tasting here ranges from Soave Classico, made from 100% Garganega, to blends with the lesser-known Trebbiano di Soave, and the signature Vulcaia Fume, a remarkable Sauvignon.

While a visit to Inama is the perfect introduction to Soave, it also offers the chance to discover different wines, predominantly reds, which are produced in the neighbouring Colli Berici – the latest buzzing wine region of the Veneto. Inama concentrates on Carmenere, Merlot and Cabernet Sauvignon, but other wineries such as Gianni Tessari are also experimenting with Pinot Nero, Tai Rosso and bubbly Durello.

www.inama.wine; tel 045 610 4343; Località Biacche 50, San Bonifacio; Mon-Fri by appointment $

04 VENISSA

You can't come to the Veneto and not visit Venice, and now there's another great reason to head here. Gianluca Bisol, an historic Prosecco winemaker, has heavily invested in replanting a micro-vineyard in the Venice lagoon, on the near-deserted Mazzorbo island. He has chosen an all-but-forgotten native grape, Dorona, and while the yields are very small, the Venissa wine is quite incredible, especially because the vineyard is almost a 'non-terroir', with sandy, salty soil at sea level.

The grapes are vinified in the nearby Colli Eugani winery of Maeli, but visitors here are treated to a tasting in the ancient wine *tenuta* (estate) and a tour of the vineyard. You can stay overnight in the small hotel, and dine in a Michelin-starred restaurant overlooking the vines.

www.venissa.it; tel 041 527 2281; Fondamenta Santa Caterina 3, Isola di Mazzorbo; daily by appointment $

05 Michelin-starred meals at Venissa's osteria

06 Displaying the produce at Le Bignele

05 MASSIMAGO

While the historic Valpolicella vineyards stretch from Verona towards the shore of Lake Garda, many younger winemakers are making their mark in the other direction, towards the rolling hills of Soave. This is certainly the case for Camilla Rossi, who took over the family estate in 2003 when she was only 20.

This rambling *tenuta* is hidden away in the middle of thick woods and olive groves. Guests can stop off for a tasting or stay over in the luxurious Wine Relais, complete with vineyard pool and spa. Camilla bubbles with enthusiasm, proclaiming, 'This is not a traditional winery. We're not interested in making wines like our *nonno* [grandad].'

Customers can create their own personalised label, and the winery regularly commissions classical music, downloadable from its site, to be enjoyed with particular wines.

www.massimago.com; tel 045 888 0143; Via Giare 21, Mezzane di Sotto; Mon–Fri

06 VALENTINA CUBI

Valentina Cubi epitomises the new generation of innovative female winemakers in Valpolicella. Arriving at her state-of-the-art winery, be prepared for some surprises. Valentina bought this 10-hectare (25-acre) vineyard in 1970, but rented it out and only took control in 2000 when she retired as the village schoolteacher.

She immediately began a production that was certified organic, rare in Valpolicella, and has launched the surprising Sin Cero, a natural Valpolicella without sulphites. Her wines reflect her personal philosophy: 'Real Valpolicella doesn't have to have pretensions, and should be light and easy to drink, good with just a plate of salami.'

www.valentinacubi.it; tel 045 770 1806; Località Casterna 60, Fumane; Mon–Sat by appointment $

07 LE BIGNELE

Visiting Luigi Aldrighetti's traditional winery is like stepping back in time. He gestures to the vines surrounding the cantina, explaining that 'for Valpolicella, we use the classic Pergola Doppia system of high hanging grapes that grow off both the right and left of the main vine – like two outstretched arms'.

The 75-year-old Luigi says he now leaves everything to his two children, Nicolo and Silvia, but he still keeps a pretty sharp eye on everything. Chattering away mainly in Venetian dialect, he opens his Classico Superiore, Ripasso, Amarone and Recioto, excellent-value vintages dating back to 2013, and says with a glint in his eye that 'the secret of a great wine here is *uva sana*, a healthy grape, rather than complicating things too much in the cantina'.

www.lebignele.it; tel 388 406 6545; Via Biniele 4, Frazione Valgatara, Marano di Valpolicella; daily $

ESSENTIAL INFORMATION

WHERE TO STAY

ALICE RELAIS NELLE VIGNE
Right at the beginning of the Prosecco road, the Cosmo brothers run an excellent winery, Bellenda, while their wives, Cinzia and Marzia, have transformed a grand mansion into a B&B. *www.alice-relais.com; tel 0438 561 173; Via Gaetano Giardino 90, Carpesica di Vittorio Veneto*

06

AGRITURISMO SAN MATTIA
This biodiverse *agriturismo* (farmstay) produces award-winning wines and grows its own vegetables for use in its restaurant. *www.agriturismosan mattia.it; tel 045 913 797; Via Santa Giuliana 2, Verona*

HOTEL TERME APOLLO
In the midst of the volcanic Colli Euganei, enjoy hot mud baths, massages and thermal pools at this family-run wellness spa, as well as a top selection of local volcanic wines. *www.termeapollo.it; tel 049 891 1677; Via S. Pio X 4, Montegrotto Terme*

WHERE TO EAT

ANTICA TRATTORIA AGNOLETTI
This historic trattoria, next door to famed Montello winemaker Serafini & Vidotto, serves such regional specialities as *risotto di radicchio rosso* and grilled *chiodini* mushrooms, plus unbelievably juicy steaks. *www.agnolettiristorante.it; tel 0422 776 009; Via della Vittoria 190, Giavera del Montello*

ENOTECA DELLA VALPOLICELLA
More than 100 Valpolicella estates are represented on the wine list of this elegant wine shop, and chef Ada pairs them with classic recipes such as *risotto all'Amarone* and pumpkin gnocchi. *www.enotecade llavalpolicella.it; tel 045 683 9146; Via Osan 45, Fumane*

WHAT TO DO

Two Palladian villas that you shouldn't miss are lovely Villa Pisani in Stra on the Riviera del Brenta; and Villa Maser, outside Asolo, which is decorated with several fabulous frescoes by Veronese. *www.villapisani.beniculturali.it; www.villadimaser.it*

CELEBRATIONS

Historic Verona is well worth visiting at any time of the year, but for a special occasion book tickets for an outdoor opera at its immense and truly spectacular Roman Arena. *www.arena.it*

01

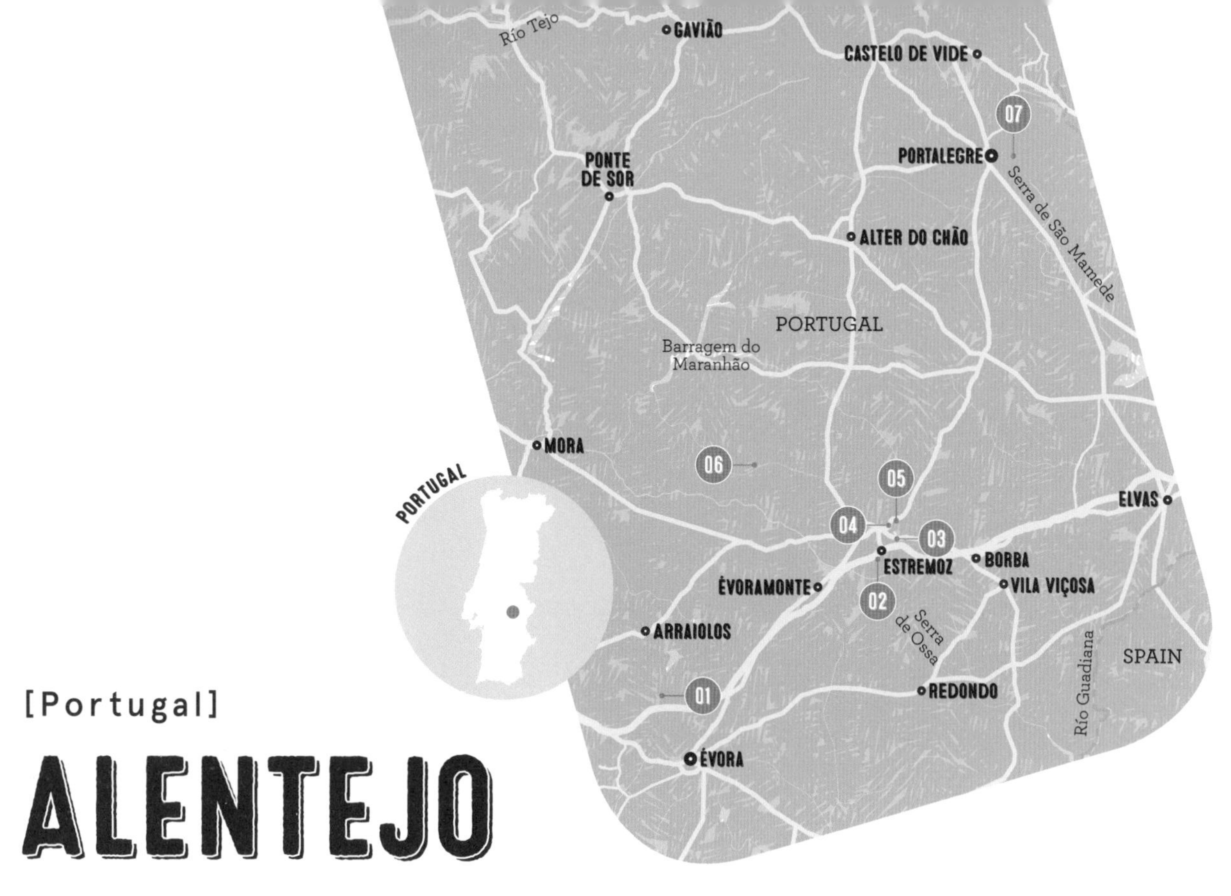

[Portugal]

ALENTEJO

With over 3000 sunshine hours a year, Alentejo's slow pace allows time to savour rich flavours strongly rooted in Mediterranean gastronomic and winemaking traditions.

Crossing the Tagus from Lisbon, you are drawn into another world: Alentejo means 'beyond' (*além*) the Tejo (the Tagus River). With vast skies and uninterrupted views, Alentejo, in Portugal's interior, offers serenity and space after the capital's bustle. Beyond the wheat fields, the rolling landscape is punctuated by olive trees and the cork and holm oaks whose acorns bring unique flavour and texture to meat from Alentejo's *porco preto* (black pig). The tinkle of bells heralds the sheep and goat herds whose milk fuels the distinctive Serpa, Nisa and Évora cheeses.

Naturally, cork is the stopper of choice for wines from Alentejo's vineyards, which cover an area the size of Belgium, with eight diverse subregions. From the atypically mountainous Portalegre in the north, the region extends south to the Algarve. To the east, approaching the Spanish border, fortified medieval towns are scattered across the hilltops.

Since the 1990s, Lisbon's proximity and the mechanisation-friendly landscape have attracted major investment in large estates (*herdade or quinta)*, with modern vineyards and state-of-the-art wineries. A wave of smooth, ripe, fruit-driven wines (mostly reds) of great international appeal has emerged, majoring on the varietals Aragonez, Trincadeira and Alicante Bouschet and, for whites, Antão Vaz, Arinto and Roupeiro. Touriga Nacional and Syrah thrive here, too. They sit alongside traditional rustic reds and an exciting fusion of the old and new following the revival of traditional grape varieties (notably in Portalegre) and techniques (for making *vinho de talha*, in particular).

Though Alentejo is large, the major highways, well-executed *agriturismo* projects and wine routes make it easy to navigate. The Alentejo wine commission's tasting room in Évora provides maps and details of its more than 60 producers.

GET THERE
Évora is 130km (81 miles) southeast of Lisbon airport; flights also arrive at Faro in the Algarve, 225km (140 miles) south of Évora.

01 FITAPRETA

A beautiful 14th-century medieval palace is home to one of Évora's largest, oldest wineries. Its turreted walls contain a surprise: a thoroughly modern, cork-clad winery. Enfant terrible António Maçanita honours the best of tradition, while embracing innovation. Classic Alentejo grapes are celebrated in style, including Antão Vaz, Roupeiro, Arinto (whites) and Aragonês (Spain's Tempranillo), Trincadeira and Alicante Bouschet. Signature Series showcases António's adventurous spirit, as do his Douro and outstanding Azores' wines, also shown here by the knowledgeable team of experts, including winemakers and sommeliers.

www.fitapreta.com; tel 918 266 993; Paço do Morgado de Oliveira, EM527 Km10, Nossa Senhora da Graça do Divor, Évora; tastings at the wine bar daily, tours and tastings twice daily by appointment

02 HERDADE DOS OUTEIROS ALTOS

'Slow food, slow wine, everything is slow in Alentejo, it's perfect,' says Jorge Cardoso. It took him and Fernanda Rodrigues three years to find the perfect spot to plant their organic vineyard, on the schist-strewn slopes of the Serra d'Ossa. Tending the vines naturally is not the only throwback to the past; they ferment and age traditional grapes on skins in *talha*. Personally hosting visits, the couple share their philosophy and welcome participation in harvest and *talha* winemaking activities. Local platters feature their own organic vegetables, herbs, olives and chutneys.

www.herdadedosouteirosaltos.pt; tel 966 250 063; Monte da Tapada Nova, Caixa Postal 11, Santa Maria, Estremoz; by appointment only, weekends best

03 TIAGO CABAÇO

Providing a stark contrast with Estremoz castle, perched above, the sleek modern lines of Tiago Cabaço's arched winery find reflection in wines labelled '.com' and 'blog', the ultra-concentrated flagship which Tiago, a millennial

01 Évoramonte

02 Fitapreta's 14th-century winery

03 The medieval interior at Fitapreta

04 Antonío Maçanita

and former teen motorcycle racer, designed himself. Made by leading winemaker Susana Esteban, his single varietal bottlings offer an unusual opportunity to get to know individual Portuguese grapes in this land of blends (including an Encruzado white, which is rarely found outside the Dão region in northern Portugal).
www.tiagocabacowinery.com; tel 268 323 233; Quinta da Berlica – Mártires, Apartado 123, Estremoz; daily by appointment, with dining option $

04 QUINTA DO MOURO

Wine-loving dentist Miguel Louro was determined to make brooding, classical reds from his small, dry-farmed estate in Estremoz, even if it meant supplementing local grapes with Cabernet Sauvignon, Merlot and Petit Syrah. The elevated location, poor schist and functional rather than fancy winery deliver spicy, structured, age-worthy reds, with a drier profile than most. Miguel, a keen hunter and antiques collector, recommends pairing these intensely flavoursome wines with local game, such as wild boar, partridge or woodcock. Top wines are foot-trodden in mills, before being basket-pressed and aged in French and Portuguese oak barrels.
www.quintadomouro.com; tel 268 334 097; Rua Antiga Estrada Nacional, Vimieiro; Mon–Fri by appointment $

05 DONA MARIA

Indolently swaying in the breeze, palm trees speak of the exotic origins of this palatial 18th-century estate, reputedly a gift from King João V to his mistress, Dona Maria. It is also known as Quinta do Carmo, after the chapel which is dedicated to Our Lady of Carmel. The current mistress of the house, Isabel Bastos, receives visitors for lunch or dinner, affording a glimpse of the house's original baroque architecture and extensive 18th-century azulejos. Following a partnership with Bordeaux kingpins Domaines Barons de Rothschild, Isabel's husband, Júlio Bastos, created their own label. The exceptionally polished range features Bordeaux grapes,

05 Fine food and wine at Herdade dos Outeiros Altos

06 Quinta do Mouro winery

but gives star billing to local hero Alicante Bouschet, which is foot-trodden in original marble *lagares* (shallow vats).

www.donamaria.pt; tel 268 339 150; Quinta do Carmo, Estremoz; daily by appointment

06 CABEÇAS DO REGUENGO

When former ballet dancer turned wine writer João Afonso acquired his quinta in Portalegre in 2009, the area had been long neglected. Today, big names are acquiring vineyards here. Why? Because this northernmost Alentejo subregion, nestled high up in the Serra de São Mamede, produces mid-weight, relatively fresh wines, thanks to higher rainfall and a cooler climate. For João, another lure was its small 'crazy' plots of gnarled, centenarian field blend vines – a mix of classic and obscure Alentejo and northern Portuguese varieties, typically interspersed with equally gnarly olive and fruit trees. Passionate about conserving this unique heritage, João tends his vineyard biodynamically and is plugging gaps (where vines have died) with his own cuttings. With an edge of wildness, the minimal-intervention wines burst with character, and some vines are over 100 years old. You won't find a more eloquent guide than João.

www.cabecasdoreguengo.com; tel 964 356 090; Estrada dos Moleiros 15, Reguengo; daily by appointment

07 HERDADE DO MOUCHÃO

Out on a limb, this historic estate was acquired in 1874 by John Reynolds for cork production. The Reynolds family's first vines were soon planted here at Mouchão and at Quinta do Carmo (see Dona Maria on p199), including Portugal's first Alicante Bouschet. Renowned in Alentejo for its colour and savoury tannins, Alicante Bouschet is a rare teinturier grape (dark skinned and fleshed), which forms the backbone of Mouchão red. It was first bottled in 1949 and Mouchão's ownership and winemaking has changed little since, save for a decade of expropriation following Portugal's 1974 revolution (look out for the barrels branded 'Co-operative 25th April', attesting to the moment when Albert 'Bouncer' Reynolds was threatened by a worker at gunpoint). Mouchão is still foot-trodden in stone *lagares* and aged for three years in old wooden *tonéis* (5000L vats), while Bagaceira, a brandy, has been made here towards the end of the harvest for over one hundred years: Alicante Bouschet marc is distilled in a copper batch still, dating from 1929.

www.mouchao.pt; tel 268 539 228; Casa Branca, Sousel; Mon–Sat, drop-in vineyard and winery visits welcome (no charge), tastings by appointment

ESSENTIAL INFORMATION

WHERE TO STAY

M'AR DE AR AQUEDUTO

This light, airy, five-star boutique spa hotel retains its 15th-century Manueline facade, but the interior is decidedly contemporary and there's also a relaxing garden and outdoor pool. Easy to find, with car parking, the hotel is tucked just inside Évora's walls and overlooks the city's imposing 16th-century Agua de Prata aqueduct.
www.mardearhotels.com; tel 266 740 700; Rua Cândido dos Reis 72, Évora

CABEÇAS DO REGUENGO

Surrounded by biodynamically farmed vines, olive trees and orchards, João Afonso's agriturismo project in the Serra de São Mamede National Park offers a bucolic bed for the night. Sheep, hens and goats keep weeds and pests at bay. Activities include harvesting grapes and olives and themed wine tastings and courses.
www.cabecasdoreguengo.com; tel 245 201 005; Estrada dos Moleiros 15, Reguengo

06

WHERE TO EAT

LUAR DE JANEIRO

This family-run place is renowned for Olivia Prates' mastery of traditional Alentejo cuisine – made with the freshest, top-quality produce – and a well-chosen wine list.
www.luardejaneiro.com; tel 266 749 114; Travessa do Janeiro 12/12A, Évora

BOTEQUIM DA MOURARIA

Tucked away in Évora's old Moorish quarter, Florabela and Domingos Canelos' restaurant offers simple, excellent home cooking. Perch at the counter on one of eight stools (no tables, no reservations, so arrive early).
Rua da Mouraria 16A, Évora

TOMBA LOBOS

Portalegre-born chef/patron José Júlio Vintém's earthy yet refined menu elevates humble local ingredients in standout dishes, such as roasted lamb, hare with white beans, and codfish rice.
Tel 245 906 111; Rua 19 de Junho 2, Portalegre

WHAT TO DO

ÉVORA

A Unesco World Heritage Site, Évora embodies Portugal's golden age. Inside 14th-century walls, winding lanes lead to striking historic buildings. Highlights include the Roman temple, the Gothic cathedral and the macabre Capela dos Ossos (bone chapel) at the church of St Francis.

WALKING

In Portalegre, stretch your legs in the Serra de São Mamede National Park, or climb the castle at Castelo de Vide, a beautifully preserved medieval town. Nearby, Coureleiros Megalithic Park features what is the Iberian Peninsula's tallest menhir – the 7m-high Menhir da Meada.

CELEBRATIONS

On 11 November, the Feira de São Martinho marks when *talha* wine is officially deemed ready to be skinned and drunk. Try it at traditional *tascas* (inns) where it's made, including País das Uvas, in Vila de Frades, and Adega da Casa de Monte Pedral in Cuba. About the same time, Herdade do Rocim hosts the Amphora Wine Day, with tastings of clay-aged wines from around the world.
www.rocim.pt

01

[Portugal]

DOURO

Take a slow boat (or train) along the beautiful Douro Valley in northern Portugal, to experience historic estates, riverside vineyards and some of the world's best reds.

Still enough to reflect the slowly shifting clouds overhead, the Douro flows westwards for more than 850km (530 miles) from central Spain (where it is called the Duero) to Porto, on the Atlantic coast. Gorge-like in parts, the rugged 100km (62-mile) valley which bears its name is the world's largest mountainside vineyard. Rising up on either side of the river, its steep, schist slopes have been carved into neat dry-stone terraces, ribbed with vines, whose leaves shimmer in the heat haze.

The whitewashed manor houses dotted among the riverbank terraces are the wine estates known as *quintas*, ranging from the grandly traditional 18th-century farmhouse of Quinta de la Pacheca to the sleek architecture of the Quinta do Seixo. On these estates, world-famous port wines, boosted by a touch of *aguardente* (distilled grape spirits similar to brandy) have been produced since the 17th century. They also make superb dry red and white wines, and rosés – the perfect showcase for the beguilingly floral perfume of grapes Touriga Nacional and Touriga Franca.

Douro winegrowing harks back to Roman times, when the arduous process of hand-carving the stone terraces began. The valiant efforts of generations past, who pruned and harvested vines and maintained terraces by hand, continue today. And although robotic 'treading machines' can emulate the process, old habits die hard. During harvest in September, when golden light bathes the valley and dust clouds billow after truckloads of grapes en route to the wineries, workers still crush grapes with their bare feet in *lagares* (shallow stone fermentation vats), accompanied by accordion music and, of course, wine. It's a beautiful time to tour the Douro.

GET THERE

The nearest airport is Porto, 127km (79 miles) from the Douro. Reachable by boat, train or car, Pinhão is the best base.

01 Terraced vines in the Douro

02 Barrel cellar at Quinta do Crasto

03 Fine dining at Quinta Nova de Nossa Senhora do Carmo

04 Dow's Quinta do Bomfim vineyards

05 The Douro near Pinhão

01 QUINTA DO CRASTO

The Roquette family run this 400-year-old estate, perched on an outcrop, overlooking the Douro River. The astounding view from the infinity pool has become almost as famous as the family's pioneering red wines. Among the region's best, old field blend vines – a mix of native varieties (49, at the last count, in top parcel Vinha Maria Teresa) – produce terrific complexity, structure and rich fruitiness (look for raspberry, cherry and blackberry) tempered by a mineral edge. They make an interesting contrast with Crasto's single varietal range (a Touriga Nacional, Touriga Franca and Tinta Roriz) and wines from a younger vineyard, Quinta da Cabreira, in the drier, warmer Douro Superior subregion. Do try the ports; unfiltered (unlike most big volume examples), the Late Bottle Vintage Port punches above its weight. *www.quintadocrasto.pt; tel 254 920 020; Gouvinhas, Sabrosa; by appointment*

02 QUINTA NOVA DE NOSSA SENHORA DO CARMO

Acquired in 1999 by the Amorim family, the world's largest cork producers, this estate's name (meaning 'new') is apt. Sprawled along a bend in the river, its south- and west-facing, sun-soaked vineyards originally produced Burmester Port. Today, they're mostly used for wine. Diverse and well executed, from easy-going Pomares to refined flagships Mirabilis white and red, the range includes a sophisticated rosé. Enjoy a glass and the view of the Douro River from the tasting room's terrace or stay longer at the 18th-century manor house's boutique hotel or highly regarded restaurant, Conceitus (BYO permitted, no corkage). Three scenic vineyard walks of varying distances take in the chapel, orchards, olive-oil mill and Marquis de Pombal milestone, one of 335 demarcating the region's original boundary. History fans will also enjoy the small wine museum. *www.quintanova.com; tel 254 730 430; Covas do Douro; daily by appointment*

Sandra Tavares da Silva and Jorge Borges Serôdio bought a warehouse in which their first batch of grapes was crushed by foot

03 QUINTA DE LA ROSA

This small, pretty Douro riverside estate was given to owner Sophia's Bergqvist's grandmother, Clara, as a christening present. Rows of vines plunge from 400m (1312ft) right down to the river, where the high, heat-reflective walls of the Vale do Inferno vineyard help explain the high-quality white and red wines and ports that are made here. Handily located within walking distance of Pinhão, the property has 23 rooms (21 with river views). Facing the river, the glass-fronted restaurant, Cozinha da Clara, also has a sizeable terrace.

www.quintadelarosa.com; tel 254 732 254; Pinhão; tours daily Apr–Oct, private visits by appointment

04 WINE & SOUL

The careers of Douro's younger generation often start by making garage wines. Sandra Tavares da Silva and Jorge Borges Serôdio had larger ambitions and bought a warehouse in the sought-after Pinhão Valley. The warehouse came with granite *lagares* in which their first batch of grapes, from a tiny plot of 70-year-old vines, was crushed by foot. This wine became the muscular, award-winning Pintas. The *terroirists* have since acquired more old field blend vineyards and artfully tease out the differences. Quinta da Manoella is slightly higher and breezier, with humid, forest influences; it produces an elegant, spicy red. The Pintas vineyard is the basis of a brooding Vintage Port, while Manoella has traditionally produced Tawny Ports, including 5G, a limited-edition Very Old Tawny Port made by Serôdio

05

P-235-AL
DOURO TAWNY

06 Quinta da Gricha's Patio das Laranjeiras

07 Churchill's Quinta da Gricha grounds

08 Stylish sleeping at Churchill's

Borges' great-great-grandfather at the end of the 19th century. Guru, a stunning, flinty white, is from even higher vineyards in Porrais. *www.wineandsoul.com; tel 254 738 076; Avenida Júlio de Freitas, Vale de Mendiz, Pinhão; Mon–Fri by appointment*

05 DOW'S

Five generations of the Symingtons have made port at Quinta do Bomfim, just a short walk from Pinhão station. Rising to almost 400m (1312ft), the estate's position is key to Dow's relatively austere style. Guided visits explore its history, supported by photographs and port wine paraphernalia from the family archive. During harvest, you can see port wine in the making from a viewing area over the robotic *lagares*. Year-round, the visit includes the 19th-century lodge, with its imposing *balseiros* (huge, upright wooden casks) and port pipes. Tastings extend to the family's other port brands, including Graham's and Warre's. Order a traditional picnic basket and enjoy it (and fine views) from the visitor centre or vineyard terraces. At the time of writing, a traditional Portuguese restaurant was scheduled to open on site. *www.symington.com; tel 254 730 370; Pinhão; daily by appointment, closed Mondays Nov–Mar*

06 CHURCHILL'S

In contrast to the other stops on this trail, Churchill's Quinta da Gricha is located on the north-facing bank of the Douro, near Ervedosa do Douro. The company was established in 1981 by Johnny Graham whose great-great-grandfather founded Graham's Port in 1820 – port clearly runs in the blood. Churchill's acquired Quinta da Gricha in 1999, and has since embarked on making Douro wines, as well as ports – all foot-trodden in traditional granite *lagares*. The single estate red wine is particularly mineral, with resinous *esteva* (a local wild plant) notes – a thumbprint which you will also find in wines from Quinta de S. José, with whom Churchill's share a jetty (boat transfers from Pinhão can be arranged). Lunch is served alfresco on the Patio das Laranjeiras (meaning 'orange trees') or in the beamed *caseiros'* (caretaker's) kitchen, with its traditional bare stone walls. *www.churchills-port.com/quinta-da-gricha; tel 254 422 136; Ervedosa do Douro, S. João da Pesqueira; daily by appointment*

ESSENTIAL INFORMATION

WHERE TO STAY

CASA DO VISCONDE DE CHANCELEIROS
Despite the grand facade, this 18th-century manor house hotel has an unpretentious, homely feel to it. Extensive views over the valley and the hotel's spectacular terraced garden (with patios, pool and tennis court) make breakfast a leisurely affair. It's a ten-minute drive from Pinhão.
www.chanceleiros.com; tel 254 730 190; Pinhão

QUINTA NOVA DE NOSSA SENHORA DO CARMO
Located on the Quinta Nova winery estate, this 19th-century manor house offers luxurious rooms with a mix of traditional furniture and modern fittings. The restaurant is exceptional, serving contemporary takes on classic Douro dishes, which are expertly paired with wines from the estate.
www.quintanova.com; tel 254 730 430; Quinta Nova, Covas do Douro

WHERE TO EAT

DOC
Famous for its superb wine list and waterfront setting (with outdoor deck), this is acclaimed local chef Rui Paula's first restaurant. DOP in Porto city centre followed in 2010, and then Casa de Chá on the beach in Leça da Palmeira, which earned Paula a Michelin star in 2016.
www.ruipaula.pt; tel 254 858 123, Estrada Nacional 222, Folgosa, Armamar

O PAPARICO
It's worth the short taxi ride north of Porto to O Paparico. Portuguese authenticity is the name of the game here, from the romantically rustic interior of stone walls, beams and white linen to the menu that sings of the seasons. Dishes such as veal with wild mushrooms and monkfish are cooked with passion, served with precision and expertly paired with wines.
www.opaparico.com; tel 225 400 548; Rua de Costa Cabral, Porto

WHAT TO DO

DOURO CRUISING
Cruise all the way from Porto and pick up a hire car in Régua or Pinhão.
www.cp.pt

MUSEU DO DOURO
Get the lowdown on the Douro Valley's history, landscape and cultural traditions, and check out the permanent exhibition on Douro and Porto wines.
www.museudodouro.pt

WALKING
A short distance north of the Douro Valley is the Parque Nacional da Peneda-Gerês, a spectacular region of mountains and forests where wild horses, boar and wolves still roam. The park is created by the folds of four mountain ranges and is the perfect place for a couple of days of hiking, punctuated by cooling dips in the clear rivers and pools.

CELEBRATIONS

Midsummer is when one of Europe's most enthusiastic street festivals takes over Porto: Festa de São João do Porto, celebrating St John the Baptist. The party starts on the afternoon of 23 June and features live music and dancing, barbecues, fireworks and a lot of wine.

ROMANIA

[Romania]

DRĂGĂȘANI

At the heart of Eastern Europe, Romania is steeped in myth and tradition, but the wine-producing region of Drăgășani pulsates with new methods and rare local grapes.

Wolves, deer and bears roam brooding Carpathian Mountain forests; and impressive castles including Bran, supposed lair of Bram Stoker's fictional count, loom on rocky hilltops. But there's more to Romania than spurious Dracula connections, from ancient monasteries, cathedral-like salt mines and a hospitable culture, to its fabulous and fascinating wines.

Romania is a wine country through and through, with grapevines grown just about everywhere. Wine may have been made here for at least 4000 years and certainly before the region became Roman Dacia; and it's said that ancient Thrace (part of today's Romania) was the birthplace of Dionysius, the god of wine. Nowadays the country is Europe's fifth biggest winegrower and produces more wine than New Zealand (though is nowhere near as famous). Travellers will find familiar grapes such as Merlot, Pinot Noir and Pinot Grigio, but more intriguing are Romania's local varieties, including the widespread Fetească group: Fetească Albă (the white maiden grape), Fetească Regală (royal maiden) and Fetească Neagra (black maiden), as well as the rare white grape Crâmpoșie Selecționată, and reds such as Novac and Negru de Drăgășani, many of which are grown nowhere else in the world.

Recent years have seen a revolution in winemaking here, with exciting wines being crafted by passionate people. This is epitomised in the southwestern Drăgășani region, likened to Romania's Tuscany for its sunny climate and gentle hills overlooking the River Olt. Wine has been renowned here since the 16th century, and today, Drăgășani is home to some of Romania's most innovative small family wineries, offering a warm and very personal welcome, and situated conveniently close together to form an easy wine trail.

GET THERE
The nearest airport is 70km (43 miles) from Drăgășani at Craiova. Bucharest has more flights but is a 2.5 to 3hr drive away (200km/124 miles).

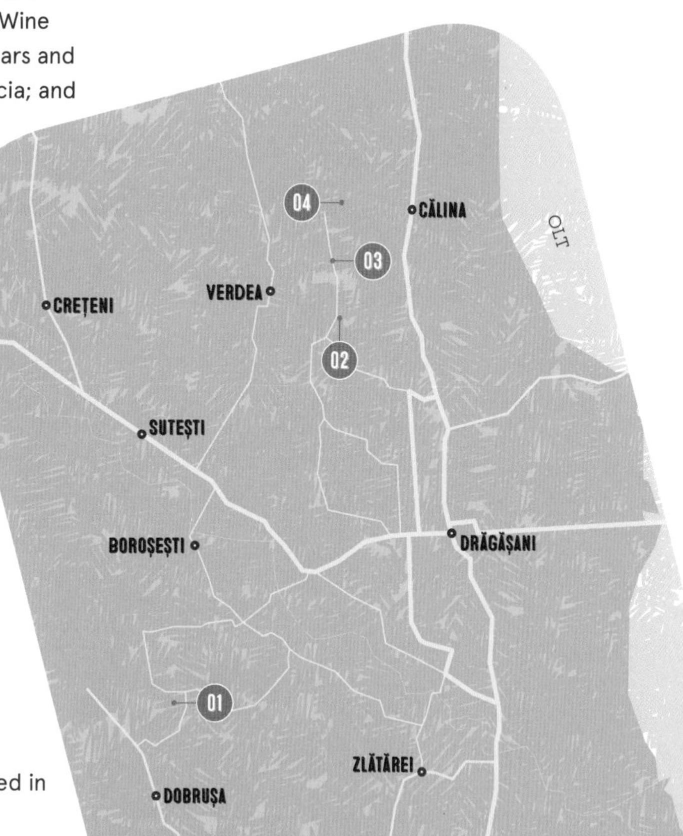

01 AVINCIS

Sitting on a plateau, surrounded by vines in every direction, this winery must be a contender for the most beautiful in Romania. The family connection dates back to 1927 when Iancu Râmniceanu (a Romanian army officer) and his wife Maria bought the Neo-Romanian mansion with vineyards. It was confiscated by the state in the communist era, but in 2007, their great-granddaughter Cristiana (together with her husband Valeriu) returned to the family estate. They restored the ruined mansion and built a dramatic modern winery in local stone with an eco-friendly grass roof to fit into the landscape. A gently sloping 40 hectares (99 acres) of vines are managed with low yields and fanatical attention to detail. The year 2011 marked the first vintage of the new era, made by a young Alsace winemaker Ghislain Moritz, who is now the winery's consultant. Local varieties are a real strength, and include the refreshingly elegant Crâmpoșie Selecționată, the fine Feteasca Regală and the wonderfully juicy red, Negru de Drăgășani.

www.avincis.ro; tel 0751 199 415; Vila Dobrușa, Valea Caselor Strada 1A; Mon–Sun by appointment

02 PRINCE ȘTIRBEY

It's not often you set foot in the home of actual blue-blooded royalty, but extending a warm welcome to visitors has always been important at Prince Știrbey. The effortlessly gracious Ileana Kripp comes from a long line of Romanian nobles – her grandmother was Princess Maria Știrbey – though the family's vineyards were seized by the communists in 1949. Ileana escaped to France aged 15 and somehow 'lost' her return ticket. She later met her husband, Jakob Kripp, in a German vineyard and, perhaps inevitably, the idea of recovering Ileana's inheritance began to play on the couple's minds. In 2001, the estate eventually returned to their hands. Luckily it had been well looked after by the state, under Dumitru Nedelut – Ileana and Jackob have kept on the same man

04

as their vineyard manager today. Their German-born winemaker Oliver Bauer has also been a key influence. He originally came to Romania for a couple of months but found the chance to rebuild a historic winery from scratch too hard to resist. From the start, the concept was unusual for Romania – with special focus on local grape varieties, such as the first Crâmpoșie varietal wine, the first dry Tămâioasă Românească, and the first Novac. Fourteen years on, Prince Știrbey is firmly established as one of the country's most consistent and exciting wineries.
www.stirbey.com; tel 0751 252 272; Dealul Olt Strada; daily by appointment

German-born Oliver Bauer found the chance to rebuild a historic Romanian winery from scratch too hard to resist.

03 CRAMA BAUER

Oliver and Raluca Bauer, who both work at Prince Știrbey, which is just 2km (1.2 miles) up the road, met at a wine tasting. They started planning their own family winery in 2010. Oliver was keen to explore Romania's winemaking potential further – his approach is to seek out rare varieties and parcels from old vineyards, creating what is effectively a single-vineyard wine from each batch. He is also conducting various winemaking experiments, such as producing orange, skin-fermented Sauvignonasse, semi-sweet Crâmpoșie Selecționată in a Mosel style, and using Negru de Drăgășani for rosé. His Fetească Neagră is excellent, too. The Bauers take a very personal approach to guiding visitors around their winery and can offer guided tastings accompanied by local foods, or special horizontal or vertical tastings.
www.cramabauer.com; tel 0757 098 940; Dealul Olt Strada; Mon-Sat by appointment

01 An aerial view of Avincis estate

02 Harvesting by hand

03 The modern winery at Avincis

04 Avincis tasting

05 Domeniul Drăgâşi's Pelerin guest rooms

06 Pelerin's vintage decor

07 Terrace views at Domeniul Drăgâşi

04 DOMENIUL DRĂGAŞI

Former lawyer Magdalena Enescu visited Drăgăşani to work on a legal case and fell in love. She subsequently took a 180-degree turn in her life and devoted herself to establishing this small wine estate. It has just 7 hectares (17 acres) of vineyards, planted in 2012, with breathtaking views over the River Olt. The vineyards' international varieties offer the chance to explore how the grapes suit the Drăgăşani landscape, with good Cabernet Franc and an unusual pink Pinot Grigio called Ramato under the Pelerin ('Pilgrim') label.

Close to both Ştirbey and Bauer, Domeniul Drăgâşi also offers five well-appointed rooms in the historic Pelerin house and makes an ideal place to stay. Leisurely days can be filled with wine tastings, wine-matching dinners based on delicious local ingredients and cooked by a professional chef, plus guided tours of the winery itself and the vineyards on request.

www.domeniul-dragasi.ro; tel 0744 337 033 or 0740 222 270; Dealul Viilor din Pruddeni; daily Mar-Oct by appointment

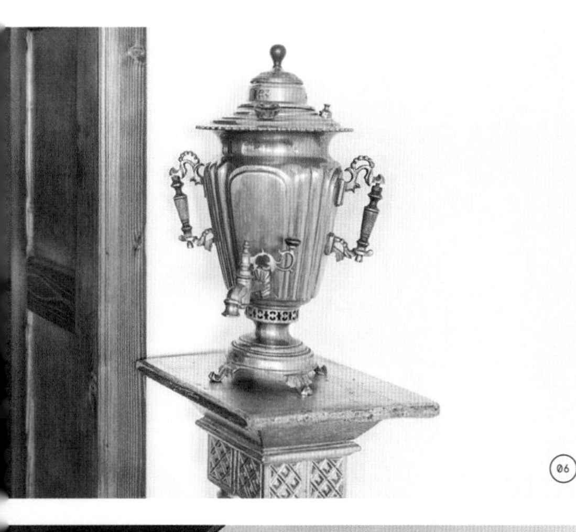
06

© Vitalie Brega | 2020. Domeniul Dragasi

ESSENTIAL INFORMATION

WHERE TO STAY AND EAT

AVINCIS

Enjoy panoramic views from one of three apartments built on the grass-covered roof of this stunning winery. There are also nine comfortable modern double rooms overlooking the vines. In addition to wine tasting and meals, you can make use of the sauna and tennis courts and walk through the vines and down to the village of Dobrușa with its small orthodox monastery. *www.avincis.ro; tel 0751 199 415; Vila Dobrușa, Valea Caselor Strada 1A*

HOTEL KMZ

In the centre of Drăgășani, this edgy, young-at-heart boutique hotel, where the 'boutique' label feels tongue-in-cheek, offers clean and cosy rooms with period design touches, such as striped bedding and gold headboards. There's a handy street-level cafe that does great sandwiches and *gyros*. *www.kmz.ro; tel 0250 814 093; Regele Carol Strada 16*

WHAT TO DO

LOCAL TRIPS

In addition to wine tastings and walks in the area, the Drăgășani winegrowers' association can help organise trips to nearby sights, such as the impressive Horezu monastery, with its Unesco Heritage-listed pottery, as well as other significant monasteries, such as Dintr-un Lemn and Arnota. It's also well worth asking about a visit to the dramatic salt mines of Ocnele Mari. *raluca@cramabauer.com; tel 0751 252 272*

WALKING

Horezu is also a good starting point for hiking and wildlife spotting in the lovely and very wild Carpathian Mountains. The Crame Romania website is a helpful resource, and also provides plenty of winery information and useful maps. *www.crameromania.ro*

01

[Slovenia]

PRIMORSKA

Sitting at the crossroads of Central Europe, Slovenia welcomes travellers with majestic Alpine peaks, wooded valleys and lakes, plus a booming wine scene.

Modern Slovenia has existed as a sovereign nation since 1991, when it seceded relatively peacefully from Yugoslavia. The history of winemaking here goes back to the 5th century BC, earlier even than in France and Italy, as vineyards were planted before the Roman era. Today, Slovenia is one of the most dynamic, exciting and innovative wine-producing regions in Europe, with the current generation of winemakers striving for top quality and offering a kaleidoscope of attractions for wine tourism: winemaker B&Bs, Mitteleuropa gastronomy paired with local wines, plus cultural and adventure side trips. Admittedly, all this comes at a cost, so be prepared for tastings setting you back around €20pp, although the wines are generally accompanied by a feast of gourmet delights produced on the *viticoltore*'s farm.

While vineyards have been planted across much of the country, three regions stand out, perfect candidates for combining on a wine-tasting voyage: Goriška Brda, the Gorizia Hills across the border from Friuli's renowned Collio; the Karst, a unique terrain and climate that follows a narrow strip of land clinging to the Adriatic up as far as Trieste; and the lush valleys and hillsides of pre-Alpine Vipava. Here you will discover many vineyards that are both certified organic and Demeter International–certified biodynamic, sample some surprising local red and white grape varieties, and get acquainted with the off-the-wall orange-wine scene. Experiencing orange wine for the first time can be a shock, the colour ranging from deep amber to dark pink, with classic grapes like Sauvignon, Chardonnay and Malvasia tasting completely different. The winemaker's trick is in macerating the skins together with the fermenting grape juice. People either love it or hate it.

GET THERE
Italy's Trieste airport is 25km (15 miles) from Dobrovo. Car hire is available.

01 Izola

02 Geurila winery

03 Primož Lavrenčič of Burja Estate

04 Sun hitting the village of Šmartno

01 KLINEC

Aleks Klinec's tiny vineyard sits on the Italian border, and the perfect place to taste his unique wines is out on the sunny cantina terrace, overlooking the rolling vine-clad hills of Goriška Brda, the Slovenian continuation of Friuli's Collio. The northern Italian influence is strongly felt in this area, in the landscape, cuisine and wine, but Aleks is a pioneer of Slovenia's renowned orange wine movement, switching entirely to traditional long skin maceration back in 2005. Signature wines are Malvasia and Ribolla Gialla, but the complex Ortodox blend, with almost 15% alcohol, also surprises with Verduzzo and Friulano grapes added too.

Aleks and his wife Simona are the consummate hosts and at weekends the winemaker cooks hearty dishes in their rustic restaurant. Pass by in October and enjoy the village arts festival when the Klinec cantina welcomes musicians, painters and sculptors from around the world.
www.klinec.si; tel 05 395 94 09; Medana 20, Dobrovo; daily by appointment

02 MOVIA

At the other end of the sleepy community of Dobrovo, charismatic winemaker Ales Kristančič has almost single-handedly brought Slovenian wine to the world stage. While vinification is undertaken in the cellar, half of his vineyards are in the Italian Collio, and some 80% of production is exported overseas. Wine enthusiasts who make it here receive a warm welcome and the chance to try biodynamic orange wines, a bubbly that is naturally fermented with no sulphite added, and if Ales is at home, a poetic explanation of his wines that 'seek the ultimate goal of purity and soul, a transparent connection to the earth'.
www.movia.si; tel 05 395 95 10; Ceglo 18, Dobrovo; daily by appointment

03 GUERILA

The first thing that impresses on a visit to the biodynamic Guerila winery is the view: the approach into the vineyard affords stupendous vistas down onto the expansive,

verdant Vipava Valley. Then you taste the wines! Zmago Petrič, who calls himself a 'nature revolutionary', only makes 40,000 bottles a year, cultivating little-known local varietals such as Zelen and Pinela, blended with Ribolla Gialla to make Castra, a sparkling Brut Nature with zero dosage. Then you taste Extreme, an experiment in amphora orange winemaking, this time using Vipavska Dolina grapes with Ribolla, where the blend is much more intense than a single grape cuvée.
www.guerila.si; tel 041 616 091; Planina 111, Ajdovščina; daily by appointment

04 BURJA ESTATE

The main Vipava Valley vineyard, Burja Estate is named after the region's famed gusty wind (the Bora, or Burja) and features 60-year-old Malvasia vines, certified organic and biodynamic. But the winery's signature white wines are all about blends, using an unexpected mix of different grapes, where Malvasia is complemented by Ribolla Gialla, Laški Rizling and Vipavec. The robust Burja Reddo blends Scioppetino, Blaufränkisch and Refosco, but the surprise is the Burja Noir, a full-bodied Pinot Noir, aged two years in big barrels, almost Burgundian in style. Sip it between mouthfuls of the homemade crusty bread and farm-cured salami with which Primož Lavrenčič welcomes visitors.
www.burjaestate.com; tel 070 900 075; Orehovica 46, Podanos; daily by appointment

05 ČOTAR

Branko Čotar originally planted this small 7-hectare (17-acre) vineyard in a quiet corner of the Vipava Valley to provide wine for the family's renowned trattoria, where it used to be served in jugs direct from the barrel. Since Branko's son Vasja joined him as the winemaker, the family still run their popular restaurant but concentrate their energies all the more on the innovative, experimental estate, specialising in volatile natural wines that have no sulphur added. White Malvasia Istriana and Vitovska have a distinctive orange colour, through the maceration of the skins during vinification. Back in the 1970s, the family excavated their own cellar chambers for barrel-ageing, and each label is signed with the winemaker's fingerprint.
www.cotar.si; tel 041 870 274; Gorjansko 4a, Komen; daily by appointment

06 ŠTOKA

The wines made by Primož Štoka and his son Tadej reflect the unique qualities found in the mineral-rich terra rossa of the windblown Karst region, roughly two to three hundred metres (660–980ft) above sea level. While the Štoka family have been tending vines here for 200 years, Primož bottled their first wine in 1989,

05 Sailing out from Izola

06 Goriška Brda

just before Slovenia broke off from Yugoslavia. Their signature vintages are barrel-aged red Teran, the local name for Refosco, and elegant, white Vitovska Grganja, an indigenous grape. The Karst is known for its cavernous limestone caves, perfect for winemakers, so don't miss a tour of the Štoka's subterranean cellar where the family cure their own prosciutto; delicious with a glass of Teran.
www.stoka.si; tel 041 667 125; Krajna vas 1d, Dutovlje; daily by appointment

07 ZARO

Matej Zaro proudly emblazons his wine labels with the family crest, claiming they have been making wine since 1348, when this part of Istria was part of the Venetian Republic. Matej's 20-hectare (46-acre) vineyard is divided up into plots around the seaside resort of Izola. The picturesque Pivol vineyards, overlooking the azure Adriatic, produce a distinctive Malvasia, used in his white blends. Rather than receiving visitors at the out-of-town cantina, Matej has transformed a 15th-century Italianate palazzo into the Manzioli Wine Bar, where wine enthusiasts can drop in for a simple tasting or try wine and food pairings.
www.vinozaro.com; tel 041 218 547; Polje 12a, Izola; daily

08 KORENIKA & MOSKON

This 26-hectare (64-acre) Istrian winery was established back in 1984, when Miran Korenika and Ignac Moškon joined forces to return the land to traditional vineyard cultivation, something which had been interrupted during the Communist era.

Start out with a taste of the easy-drinking wines, such as the crisp, white Pinot Blanc blend and the red Refosco, then move on to the more interesting Cru vintages, especially Sulne, which is made predominantly using the native Malvasia Istriana grape, which spends some six years in oak barrels.

The whole estate is certified Demeter – organic and biodynamic – and also produces its own olive oil and grape juice, as well as a very smooth grappa infused with fennel and lavender.
www.korenikamoskon.si; tel 041 854 186; Korte 115c, Izola; daily by appointment

ESSENTIAL INFORMATION

WHERE TO STAY

B&B KLINEC PLESIVO

Located right in the heart of the Goriška Brda vineyards, across town from the Klinec winery, winemaker Aleks' brother Uroš runs a stylish modern guesthouse as well as a prosciutto cellar, where he cures some very flavoursome Mangulica hams and salami. *www.klinecplesivo.si/wordpress; tel 031 339 463; Plešivo 51a, Dobrovo*

06

MAJERIJA

Originally dating back to 1700 and surrounded by verdant orchards and vineyards, the sprawling Majerija manor house makes an ideal base from which to explore the vineyards of the Vipava Valley. The ten elegant rooms within look out over a herb garden, and the sophisticated restaurant offers up a contemporary twist on the best of traditional Slovenian cuisine. *www.majerija.si; tel 05 368 50 10; Slap 18, Vipava*

HOTEL DOBREGA TERANA

Located on Slovenia's Karst, Hotel Dobrega is not actually a hotel at all, but a wine association, organising homestays with local farmers, cheesemakers, artisans and winemakers. *www.hoteldobregaterana.si; tel 031 323 191; Sejmiška ulica 1A, Sežana*

WHERE TO EAT

ARKADE CIGOJ

The friendly Cigoj family make their own Vipava wines, cultivate a herb and vegetable garden, and raise livestock, all of which end up at their farm-to-table restaurant. The ever-changing menu may feature barley and wild mushroom risotto, gnocchi with radicchio and sausage, and tempting apple and blueberry strudel. *www.arkade-cigoj.com; tel 05 366 60 09; Črniče 91, Črniče*

ČEBRON FAMILY ESTATE

Featuring tables set on a shady terrace overlooking the Vipava hills, this working farm and vineyard specialises in seasonal ingredients as the foundation for minestrone or pumpkin soup, pasta with wild boar ragu, or even olive-oil ice cream. *www.cebron.eu; tel 041 582 051; Preserje 59, Branik*

WHAT TO DO

In the Slovenian capital of Ljubljana, take a lazy river cruise or board the funicular up to the 15th-century castle. For a break from the grape, sample the ales of Union Brewery, which serves both vegetarian dishes and burgers. *www.union-pivnica.si*

CELEBRATIONS

In the heart of the Slovenian Collio, Dobrovo hosts a colourful Cherry Festival on the first Saturday of June, with parades, a food festival and wine tastings, while the Vipava Valley holds a wine and culinary festival on the second Sunday and Monday of May.

01

[Spain]

EL BIERZO

Follow in the footsteps of Romans, Templars and pilgrims to these scenic Spanish valleys to discover the dark-fruit aromas of Mencía reds and the seductive minerality of relatively unknown whites.

The fertile uplands of El Bierzo, just west of León, are where the high *meseta* plateau of central Spain gives way to the green valleys of Galicia, the country's far northwestern corner. El Bierzo is not just a wine appellation, but a separate locale with its own culture and a very distinct feel from the rest of the huge Castilla y León region.

It was the Romans who first brought viticulture to the area, when they developed El Bierzo as a mining outpost. Their legacy can still be found in place names and the tortured rocks of Las Médulas, once a Roman goldmine. Celtic ruins, ancient churches and a Templar castle are among many other historic sights here.

You'll also see the Camino de Santiago's scallop shell emblem everywhere: the Camino Francés – the most popular of the famous pilgrimage's various routes – passes through El Bierzo, and the region's wine has been the comfort of pilgrims since medieval times. As has the hearty, porky *berciano* cuisine, with its excellent *embutidos* (cured meats), stuffed peppers, warming *cocido* stews and signature *botillo* (El Bierzo's answer to haggis), all of which accompanies El Bierzo's red wines very well.

GET THERE
Trains run to Ponferrada, buses to other towns such as Villafranca del Bierzo. You'll need a car to reach rural bases.

These are traditionally made from the Mencía grape, which has a characteristic dark cherry aroma. There are also white wines, with Godello a particularly interesting varietal with a complex mineral finish. The wine region attracted more attention in the early 2000s, but the best Bierzo drops still offer excellent value for their quality. Production is from small parcels of land, rich in old vines.

El Bierzo is also deservedly proud of its other natural produce: its scenic fertility, in contrast to the parched plains of many Spanish wine regions, gives the land a more northern-European feel, with chestnuts, cherries and apples.

FABRERO
PEDRAFITA DO CEBREIRO
O CEBREIRO
FRESNEDO
06
05
Río Sil
07
SEOANE DO COUREL
03
04
BEMBIBRE
TORRE DEL BIERZO
VILLAFRANCA DEL BIERZO
PIEROS
CACABELOS
Embalse de Bárcena
VILELA
CAMPONARAYA
a do Courel
01
02
TORAL DE LOS VADOS
PONFERRADA
Río Sil
LAS MÉDULAS

01 DESCENDIENTES DE J PALACIOS

The friendly uncle-nephew team behind some of El Bierzo's best wines may be more interested in winemaking than marketing, but they nevertheless have a very striking winery south of Villafranca del Bierzo. The building is designed by famed Navarran architect Rafael Moneo and is a fitting home for reds that have won renown among connoisseurs around the world.

Alvaro and Ricardo (known as Titín to all) come from Rioja winemaking stock but have been resident in El Bierzo for over 20 years. Their best-known wine, Pétalos, offers exceptional value and is a complex, balanced drop that is nearly all Mencía. While Pétalos is sourced from various vineyards, many of their other wines are single-vineyard gems, rising in quality to the supreme La Faraona, produced from a parcel of land that only yields enough for a thousand or so bottles. Many consider it the region's finest wine. *info@djpalacios.es; tel 987 540 821; Avenida Chao do Pando 1, Corullón; Mon–Fri, mornings only*

02 PEIQUE

Totally family-run, this winery has a picturesque setting in the heart of the region. A visit includes a stroll around the vines and offers interesting information about the characteristics of the Bierzo *denominación de origen*. Peique's low-priced entry-level red is a favourite in bars around the province and offers a good introduction to the Mencía grape. Peique also make a good Godello white, a more complex red made from parcels of old vines, and a deeply coloured Garnacha Tintorera, a varietal known outside Spain as Alicante Bouschet. You are guaranteed a genuine welcome. Better to book ahead. *www.bodegaspeique.com; tel 987 562 044; Calle El Bierzo 2, Valtuille de Abajo; Mon–Fri & Sat morning* $

03 LOSADA

Housed in a handsome building inspired by nearby Roman ruins, this winery is on the road

01 Ponferrada

02 Roman bridge, Molinaseca

03 Templar castle, Ponferrada

between Cacabelos and Villafranca del Bierzo. Its wines eschew the upfront fruit of many local producers in favour of a more elegant, balanced style that brings out the best from high-quality Mencía grapes, sourced from numerous different plots around the region. There are several grades of red, with the standard-bearer Losada wine a great buy at around €12. La Bienquerida, sourced from centenarian vines, has a darker complexity to it. Losada also produces a Godello white and a rosé.
www.losadavinosdefinca.com; tel 987 548 053; Carretera LE-713, km12, Pieros; Mon-Fri (plus Sat for groups), phone to confirm $

04 CUATRO PASOS

The Galician winery Martín Codax, known for its fine Albariños, is behind this slick operation set in a handsome mansion in Cacabelos. Brown bears are known to roam around El Bierzo, and paw prints, found after a hungry specimen enjoyed a bunch of grapes in one of its vineyards, are the winery's cute signature device. Cuatro Pasos is set up for tourism more than many local wineries, and visitors have several different types of tours available, some of which must be pre-booked. It's worth upgrading from the standard 40-minute visit to one that will allow you to taste some of the premium wines, which are classy oak-aged Mencías sourced from old vines. The entry-level Cuatro Pasos is an uncomplicated but tasty quaffing red; there's also a rosé version.
www.cuatropasos.es; tel 987 548 089; Calle Santa María 43, Cacabelos; Tue-Sat $

05 PITTACUM

By the church in the village of Arganza del Bierzo, this beautiful slate building hosts a welcoming *bodega* that is one of the longer-established of the new wave of Bierzo wineries, having been here since the 1990s. The Pittacum red, made from 100% Mencía, is a powerful but balanced burst of dark-fruit flavour and is a favourite on restaurant menus through the province. It pairs well with the heartier local dishes. Tours are well-organised and engaging, with English spoken; book 48 hours in advance via the website. Pittacum's more upmarket wines are also available to taste for an extra fee.
www.terrasgauda.com; tel 987 548 054; Calle La Iglesia 11, Arganza del Bierzo; daily $

06 PALACIO DE CANEDO

The larger-than-life figure of José Luis Prada has been promoting El Bierzo since the 1990s, putting his name and face to his own wine and food label, as well as a series of franchise restaurants showcasing the best of the region's produce. Don't let his fun-loving style fool you, however – the produce is of undeniable quality.

This sizeable complex has a restaurant, a hotel and a shop as well as the winery. Though three

04 El Bierzo vineyards

05 Losada wine

06 Plate of tapas

different tour types are offered, they are essentially the same with varied tasting options. As well as visiting the *bodega*, you'll be taken for a spin through the vineyard onboard an oversized golf buggy. Wine has been made here since the 18th century.
www.pradaatope.es; tel 987 563 366; Calle Iglesia s/n, Canedo; daily; book ahead to guarantee a tour

07 DOMINIO DE TARES

In the eastern reaches of the Bierzo region, in a rather unlovely zone of warehouses, this winery is nevertheless definitely worth the stop. It combines modern techniques and tradition to advantage and there isn't a bad bottle in the range. The mid-range Mencía Cepas Viejas has deservedly won plaudits from wine writers who admire its characteristic dark-fruit acidity combined with elegant structure. P3 is the top of the range, harvested from a small patch of centenarian vines, and an excellent Godello is also produced. Reserve visits in advance, as there isn't a structured visitor experience. However, you are likely to be able to taste something if you just drop by.
www.dominiodetares.com; tel 987 514 550, Calle los Barredos 4, San Román de Bembibre; Mon–Fri

ESSENTIAL INFORMATION

WHERE TO STAY

PARADOR DE VILLAFRANCA DEL BIERZO
At the entrance to handsome Villafranca, this modern example of the emblematic state-run parador chain makes for a very comfortable base. *www.parador.es; tel 987 540 175; Avenida Calvo Sotelo 28, Villafranca del Bierzo*

HOTEL VILLA DE CACABELOS
Offering a genuine welcome and some good-value midrange comfort, this is easily the best hotel in the wine village of Cacabelos. It's very handy for numerous wineries. *www.hotelvilladecacabelos.es; tel 987 548 148; Avenida Constitución 12, Cacabelos*

HOTEL BIERZO PLAZA
Right on the square in Ponferrada, this place offers substantial comfort at a very reasonable price. *www.aroihoteles.com; tel 987 409 001; Plaza Ayuntamiento 4, Ponferrada*

WHERE TO EAT

León is a free-tapas province; expect a little snack with every glass of wine you try in a bar.

LA CASONA
Northwest of Ponferrada, in a stalwart historic mansion, this restaurant is a top place for regional *berciano* cuisine. *www.restaurantelacasona.com; tel 987 455 358; Calle Real 72, Ponferrada*

MESÓN DON NACHO
In Villafranca del Bierzo, this very traditional *mesón* does hearty local classics at very good prices. *tel 987 540 076; Calle Troqueles s/n, Villafranca del Bierzo*

WHAT TO DO

HIKING
The Camino de Santiago tracks across the region, and it's easy to tackle a stage or two. But there are lots of other well-marked hikes in the area, taking you into scenic corners away from the steady pilgrim flow. Try walks around the gorgeous slate village of Peñalba de Santiago.

LAS MÉDULAS
In a search for gold here, the Romans blew whole mountains apart in an act of environmental vandalism enabled by clever engineering. What's left is a captivating series of orange sandstone outcrops; interesting guided visits stroll you through the landscape.

PONFERRADA
The region's capital is a vibrant town where an attractive historic centre is protected by a superb Templar castle overlooking the river.

MOLINASECA
Just outside Ponferrada, this charming restored village is known for its cured meats, which you can try in the rustic taverns that are tucked into its traditional slate houses. The river through the town is a focal point of a major August water fight during the local fiestas.

CELEBRATIONS

FERIA DEL VINO DEL BIERZO
The region's food and wine are celebrated over three days in April in the village of Cacabelos. *www.facebook.com/feriavinobierzo*

FESTIVAL ESTIVAL DEMENCIAL
Held in July, this wine fair brings producers, wine critics and performers to Ponferrada's Templar castle for a fun day of tasting and music. *http://fed.delbierzo.es*

01

[Spain]

JEREZ

Sherry is stylish again. Head to the source of the ultimate tapas companion in this handsome Andalucían city for the lowdown on Fino.

Beware: you could get lost for days, maybe weeks in the Sherry Triangle. Formed by the three towns of Jerez de la Frontera (known as Sheris in medieval times), Sanlúcar de Barrameda to the west and El Puerto de Santa Mariá, this corner of Spain's southern region of Andalucía is the only source of sherry, the fortified wine that is regaining well-deserved favour among food-lovers.

Every evening in bars right across Andalucía a pre-dinner ritual is repeated. Patrons take a seat and the bartender pours a small glass of pale gold liquid, the colour of an eagle's eye, then slides across the counter a plate of *jamón ibérico*, the air-dried local ham, or some cubes of cheese. The drink is Fino, the palest and driest of sherries, alive with a mouth-puckeringly savoury tang and, as legions of bar-hoppers are discovering, it's the perfect companion to sociable snacking.

The world of sherry is small but it can be bafflingly complex – it's not one drink but about five or six, and it's not immediately obvious how a refined Fino relates to the sickly, orange sherry from your great-aunt's drinks cabinet. This wine trail will set you straight.

Sherry's journey back to the bar-top began in the mid-1990s when inland Jerez, El Puerto de Santa María and Sanlúcar de Barrameda on the Spanish coast gained recognition and protection from the European Union as Protected Designation of Origin (PDO). Your journey starts in Jerez, where all the big producers have a bodega that's open to the public. It can be quite a commercial experience, but it's a good introduction. Afterwards, take to Jerez's paved streets to get a flavour of Andalucían flamenco, then hit the bars to test your new-found know-how. The Wine Trail continues out of the city to the coast; make sure you find time to stop in fabulous, dissolute Cádiz along the way.

GET THERE
Jerez airport has a few flights from the UK. Or fly into Málaga or Seville, then drive.

02

03

01 GONZÁLEZ BYASS

Close to Jerez's cathedral, the bodega of González Byass is the city's most visitor-oriented. Fittingly, its Tio Pepe Fino is the world's biggest-selling sherry. It's the place to come to get a grip on the basics of sherry production. There are several different types, from dry to sweet. At the driest end of the spectrum (and these have led sherry's revival) are Fino and Manzanilla. The tangy flavours of Fino are caused by a yeast known as *flor* that forms a film on the surface of sherries as they rest in the barrel for a minimum of three years (often much longer for the high-end vintages). This layer protects the sherry from the air (keeping it pale in the process). After Fino, the next step for a sherry is the bone-dry Amontillado, a Fino that has continued ageing in contact with the air. As a result, it's darker and richer; expect woody and dry citrus flavours. The González Byass Amontillado is an attention-grabbing treat, and so are most of the bodega's older sherries.

Next on the sherry sweetness scale is Oloroso, which is fortified after fermentation to stop the *flor* forming. As it ages in the barrel, in contact with air, it grows darker, richer and fruitier. Pedro Ximénez is a dessert wine made from the grape of the same name; the more mature the better. Any extra-sweetened sherries beyond this point are for export only...

www.bodegastiopepe.com; tel 956 357 016; C/Manuel María González 12, Jerez; daily $

02 SANDEMAN

Harvey, Osborne and Sandeman: the names that reveal sherry's genesis. Sherry's character might be Spanish but the business is British. Blame Sir Francis Drake for Britain's sherry obsession: the Elizabethan privateer sacked Cádiz in 1587 and made off with 3000 barrels of the local vino. Before long the Brits back home had developed a taste for Spain's fortified wine and a new industry was born. Entrepreneurs such as George Sandeman from Perth, Scotland set up businesses in Jerez and the rest, as they say, is history. The Sandeman bodega was

01 José Blandino of Bodegas Tradición

02 Grapes destined for the sherry bottle

03 Tasting at Bodegas Tradición

04 Delgado Zuleta Manzanilla

05 Bodegas Tradición

04

05

'When we start out, the wines are like little children. We have to teach them how to grow so that they can become adults we can be proud of.'

–José 'Pepe' Blandino, cellar master at Bodegas Tradición

established in 1790, close to the Royal Andalusian School of Equestrian Art in the heart of the city. With guided tours in multiple languages and a museum, it's a good place to get the background of sherry's story (trivia: Sandeman's logo, the dashing, caped figure, was designed by the Scottish artist George Massiot Brown). Three tours are offered, each ending with a tasting of several sherries. ***www.sandeman.com; tel 675 647 177; C/Pizarro 10, Jerez; by appointment*** $

03 BODEGAS TRADICIÓN

In the cellar of Bodegas Tradición – boutique producer of rare aged sherries – amid the gloom of 625L casks of American oak, the smell of sherry is overwhelming. In the tasting room, among artworks by Goya and Velázquez and ceramic tiles painted by an eight-year-old Picasso, visitors seek out the unusual Palo Cortado, a nutty, smoky style lying somewhere between an Amontillado and Oloroso. The bodega's own full-bodied Oloroso combines vanilla, ginger, and the smell of Christmas cake.

Sherry, more than any other wine, requires human intervention at every step, and José Blandino, the cellar master (or *capataz)* at Bodegas Tradición, who has worked in the industry for almost five decades, treats his sherries like his own offspring. 'When we start out, the wines are like little

06 Osborne Mora winery, El Puerto de Santa María

07 Iberian lynx

08 Cathedral of San Salvador, Jerez de la Frontera

children. We have to teach them how to grow, to help them through the varying stages of getting older. It takes a lot of time and hard work, so that they can become adults we can be proud of.'

But even José admits that each person's response to the final product is as important, and as personal as his own role in the process. 'We can show people what to look for. But the only standard that really matters is whether or not you like it.'

www.bodegastradicion.es; tel 956 16 86 28; Plaza Cordobeses 3, Jerez; Mon–Fri & Sat mornings by appointment

04 OSBORNE MORA

You'll recognise the Osborne bull: that black silhouette that glowers from numerous Spanish roadsides, a stroke of advertising genius from a 1956 marketing campaign by the company (the original billboards had 'Veteran Brandy' emblazoned across them). The full story is revealed in a new gallery at Osborne Mora's winery in El Puerto de Santa María.

This recently renovated bodega, sitting just 300m (985ft) from the sea of the Costa de la Luz, is the most southern point of the Sherry Triangle. The company was started by an Englishman, Thomas Osborne Mann, in 1772 but has since added Cinco Jotas *jamón ibérico* from acorn-fed black Iberian pigs and olive oil to its portfolio – meaning that it's an ideal location for a ham-and-sherry tasting, one of the options offered.

Visitors can also learn about the *solera* system for blending sherries, in which wine of varying ages is decanted from one barrel to another, keeping barrels filled with younger wine that feeds the *flor* – this is what the stack of barrels in the cellar is all about. The company's collection of VORS (Very Old and Rare Sherries) is available for tasting (on the €40 tour). Its rare sherries include wines drawn from a single *solera*, such as the Antonio Osborne Solera, started in 1903 to mark the birth of the second Count of Osborne's son. Sherry ageing is only an average; a sherry from a *solera* started ten years ago will have ten-year-old wine in it but also younger wine. Figure it out over some *jamón* and a glass of the classy Fino Quinta in the new tapas tavern.

www.bodegas-osborne.com; tel 956 869 100; C/los Moros 7, El Puerto de Santa María; by appointment

05 DELGADO ZULETA

It's a tough gig being a sherry grape. The three varieties permitted in the production of sherry are Palomino, Moscatel and Pedro Ximénez; all have to cope with summer temperatures touching 40°C (104°F), their metres of roots driving deep down into the *albariza* soil – limestone – in the search for water. At Delgado Zuleta, since 1744 those grapes have been turned into Manzanilla, a speciality of Sanlúcar de Barrameda, where the *flor* grows thickest and the saline tang is enhanced by the cool seaside climate. A basic tour of the bodega takes in the winery and the vineyard plus a tasting of three to five sherries (from €8). Delgado Zuleta's best-known Manzanilla is La Goya, first launched in 1918 and still a deliciously savoury sidekick to the local *langostinos* (prawns).

www.delgadozuleta.com; tel 956 360 543; Avendia Rocío Jurado, Sanlúcar de Barrameda; Mon–Sat by appointment

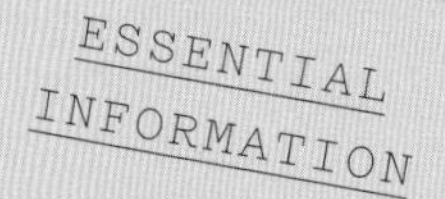

WHERE TO STAY

HOTEL CASA GRANDE

This hotel occupies a carefully restored 1920s mansion, with rooms spread over three floors and set around a patio or beside the roof terrace, which has good views of Jerez's rooftops. *www.hotelcasagrandejerez.com; tel 956 345 070; Plaza de las Angustias 3, Jerez*

LA ALCOBA DEL AGUA

Modern, central and stylish, La Alcoba del Agua is the place to book for a night or two in Sanlúcar de Barrameda at the end of the trail. *www.laalcobadelagua.com; tel 956 383 109; C/Alcoba 26, Sanlúcar de Barrameda*

WHERE TO EAT

LA CARBONÁ

Dishes at this family-run restaurant in Jerez are paired with different sherries. Inventive plates feature the best of Spanish ingredients, from spicy *chistorra* sausage from Navarra to langoustines from Sanlúcar de Barrameda. It's a trick that works well for both partners; sherry cooking classes with chef Javier Muñoz are available if you want to learn more. *www.lacarbona.com; tel 956 347 475; C/San Francisco de Paula 2, Jerez*

WHAT TO DO

DOÑANA NATIONAL PARK

One of Europe's great wildlife reserves covers the marshes, lagoons and scrub of the Guadalquivir River's estuary. Birdlife is the big draw, with half a million birds wintering here. But the Unesco-protected park is also home to the rare and threatened Iberian lynx (though you would still have to be lucky to see one). Wildlife-spotting visits must be booked in advance.

CÁDIZ

Spend a day in shabby-chic Cádiz, one of Europe's oldest continuously inhabited settlements. Romantic, mysterious and much-contested over the ages, Cádiz intoxicates with its edgy aura. *www.cadizturismo.com*

CELEBRATIONS

FIESTA DE LA VENDIMIA

Most winegrowing regions of Spain celebrate the annual grape harvest; in Jerez it kicks off at the end of August or in early September. The two-week festival opens with a ceremonial grape crush, or *pisá*, and a blessing in Jerez de la Frontera's cathedral. There follows two weeks of exhibitions, fireworks and food and wine tastings, with extra wineries and vineyards throwing open their doors to visitors.

CADIZ CARNAVAL

No other Spanish city celebrates Carnaval with as much fervour as Cádiz, which is overtaken by a ten-day singing, dancing and drinking fancy-dress party spanning two weeks in February.

07

08

[Spain]

MALLORCA

Sun, sea, sand and surprisingly good vino: head inland and discover this holiday isle's wines, made in the shadow of the Serra de Tramuntana and on the central plains.

Mallorca, the largest of Spain's Balearic Islands, has been a wildly popular holiday destination for millions of northern Europeans for almost a century – but it has survived this invasion and prospered. A group of earlier invaders – the Romans – were evicted from the Mediterranean island in AD 425, leaving behind two things: vines and olive trees. A fair price, perhaps.

Since wine has been part of the landscape of this enchanting island for more than 2500 years you can expect some old vines; grapes are routinely harvested from 60-year-old plants in Mallorca's two *Denominació d'Origen* (DO) zones, Binissalem and Pla i Llevant (the central plain). Most Mallorcan bodegas (anything from a warehouse to a stone cellar) concentrate on the island's traditional grape varieties: Callet, Mantonegro and Prensal Blanc. These are typically pepped up with dashes of Cabernet, Merlot or Shiraz. Malvasia, an ancient grape, is making a comeback on the terraced vineyards of the Serra de Tramuntana mountain range, which runs along the west edge of the island and is the first land to meet incoming weather fronts.

Wine touring on Mallorca has two great joys: firstly, most of the bodegas are within a short drive of one another and you're never more than a half an hour from a good restaurant, a beach or a sensational view. And, secondly, the vast majority of the island's bodegas are family-run businesses, often located in old sandstone buildings in the centre of towns. If you meet the families who make the wine you'll catch some of their passion for this place and its food, wine, traditions and landscapes. The island has long produced rough-and-ready wines – dark and unfiltered from sun-baked vineyards – but with the combination of decades of tradition and the ambition of younger generations, Mallorcan wine is becoming more sophisticated. Only minuscule quantities are exported, which is all the more reason to investigate them in person.

GET THERE
The island capital, Palma, receives flights from all over Europe. There's also a slow ferry from mainland Spain.

01 Mallorcan coastline

02 Modern production, Bodegas Miquel Oliver

03 Tasting at Bodegas Miquel Oliver

04 Bodegas Ribas

01 BODEGAS MACIÀ BATLE

'People thought we were crazy investing in wine rather than a golf course or hotel,' says Ramón Servalls i Batle, director of Macià Batle. 'But they're realising that there's so much more to Mallorca than sun and sand.' As one of the largest and most diversified wineries on the island – its shop stocks olive oils, chocolates and appetisers – Macià Batle stands at the opposite end of the winemaking spectrum to Miquel Oliver (see opposite). Here wines are blended in a high-tech laboratory (you can watch from behind a glass partition) and bottled by a €500,000 machine.

But for all the trappings of big business, Macià Batle remains a family-owned winery with a friendly welcome for all and a firm attachment to tradition. All Ramón's wines are based on indigenous grape varieties, such as Callet and Mantonegro, hand-picked from vines that can be more than 40 years old – in fact, some vines are so old that they produce just one bunch of grapes. The bodega has been on this site, in the foothills of the Serra de Tramuntana, just south of the wine-producing hub of Binissalem, since 1856 and the vines are managed with local know-how. 'Days are so warm that the grapes need a cooling breeze, so we take away leaves, letting the wind get to the grapes,' says Ramón. After pressing, some of the juice is matured in the crimson-painted cellars under the courtyard, where there's space for 850 barrels. A wide range of wines emerge, including a raspberry-scented *rosada*, a Blanc de Blancs and a spicy Crianza.

www.maciabatle.com; tel 971 140 014; Camí de Coanegra, Santa Maria del Camí; Mon–Sat by appointment $

02 BODEGA RIBAS

Head north towards the foothills of the Tramuntana range to explore the oldest winery on the island: Bodega Ribas. Construction began on the winery at this grand country mansion in 1711. 'Wine was part of everyday life then and the consumption was much higher than nowadays,' explains Araceli Server

Ribas. 'Many families had a plot for their own consumption and it was part of the Mediterranean diet; even kids had it regularly mixed with water.' Today, Ribas' wines are powerful and best paired with the robust Mallorcan cuisine. The rare Ribas de Cabrera and the Ribas range blend Mantonegro grapes with Cabernet Sauvignon and Shiraz. 'The Mantonegro lends a fresh nose that complements the warmer mouth,' says Araceli. Unusually, a sweet wine, Sioneta, using Muscat grapes that have dried on the vine, is also made here. Tours are offered with tapas, food or musical accompaniment.
www.bodegaribas.com; tel 971 622 673; Carrer de Muntanya 2, Consell; Mon–Fri by appointment

03 BODEGA BINIAGUAL

If you want to get fit, work in a winery. From May to September workers patrol Bodega Biniagual's 148,000 vines three times a week, walking 10km (6 miles) each time. Everything is done with extra care here because the winery has been given a second lease of life. The family-run finca (estate), set around a handsome courtyard deep in the DO Binissalem, was a farming homestead in the days of Arab rule. Set at the crossroads of the island, its golden age was in the 17th century when new houses were built and vines planted. But come the 20th century, phylloxera destroyed the vines and the community. Biniagual's resurrection began in 1999 when vines were planted once again.

Over the last two decades Mallorcan winemakers have learned how to make indigenous grape varieties the stars of authentic Mallorcan wines. 'You'll only find Mantonegro on the island, especially around Binissalem,' says winemaker José Luis Seguí. 'It requires a lot of work but gives the wines their personality and unique aroma.' In 2019 a new cellar door, hosted by sommelier José García, was opened in which to try Biniagual's wines (especially the Finca Biniagual Mantonegro), with morsels of cheese and ham.
www.bodegabiniagual.com; tel 971 511 524; Camí de Muro 11, Binissalem; Mon–Fri by appointment

04 BODEGAS MIQUEL OLIVER

This is a hard-working bodega led by the serious but enthusiastic winemakers Pilar Oliver and Jaume Olivella. You've now crossed over into the second of Mallorca's two Protected Designations of Origin (PDO): Vins des Pla i Llevant. This designation covers the terracotta plains of central Mallorca, where Miquel Oliver grows its Callet, Mantonegro and Prensal Blanc vines, divided by drystone walls from fields of almond trees (the trees blossom in February, making

the month a spectacular time to visit inland Mallorca). Pilar and Jaume opened a new cellar door next to their vineyards in 2015, to mark the winery's 2012 centenary. The original winery in the centre of Petra has been restored and now offers an insight into the island's wine traditions – tours take in both venues before Pilar and Jaume show off their most successful creations: Ses Ferritges, a Callet, Shiraz, Merlot and Cabernet blend; and Aia, an award-winning Merlot.

Petra's claim to fame is that it is the birthplace of Father Junípero Serra, the priest who not only founded the Mexican and Californian missions that became San Diego, Santa Barbara and San Francisco but also introduced vines to California. It is strange to think that Napa Valley's origins lie in a sleepy Mallorcan town that has changed little since Serra started his walkabout in 1749. Before leaving, catch the views from his hilltop hermitage, 4km (2.5 miles) out of town.

www.miqueloliver.com; tel 971 561 117; Carretera Petra-Santa Margalida km1.8, Petra; Mon-Fri & Sat mornings by appointment

05 VINS MIQUEL GELABERT

End the trail with some of the most garlanded Mallorcan wines, made on the east side of the island in a rustic estate just outside the workaday town of Manacor. From around 30 varieties of grape, some from old vines of 40 to 65 years of age, wine magician Miquel Gelabert conjures fascinating wines that are aged in a cellar dating from 1909. Notably, he produces stellar white wines, slightly unusually on Mallorca, including the excellent Sa Vall Selecció Privada, blended from Giró Blanc, Viognier and Pinot Noir.

You're only a short hop from the quiet beaches of Mallorca's east coast, so end the trip with a chilled bottle, a bite to eat and the waves breaking on the sand.

www.vinsmiquelgelabert.com; tel 971 821 444; Carrer d'en Salas 50, Manacor; by appointment

05 Bodegas Miquel Oliver wines

06 Port de Sóller

ESSENTIAL INFORMATION

WHERE TO STAY

HOTEL ES RECÓ DE RANDA
At the foot of the Puig de Randa, Es Recó de Randa is the best of the hotels in the Pla i Llevant region, close to Felanitx, Porreres and Petra. The kitchen prides itself on serving up traditional Mallorcan cuisine.
www.esrecoderanda.com; tel 971 660 997; C/Font 21, Randa

SA TORRE
Sa Torre is a very comfortable, midrange rural hotel with two swimming pools, bicycle storage and its own winery. Lying in the heart of Mallorca's wine country, just south of Binissalem, it has a highly regarded restaurant with a great collection of local wines.
www.sa-torre.com; tel 971 144 011; Carretera Santa Maria-Sencelles km7, Santa Eugènia

WHERE TO EAT

RESTAURANT MARC FOSH
One of Mallorca's most renowned chefs, Marc Fosh has won a Michelin star for his restaurant in Palma's Hotel Convent de la Missió. Kitchen vegetables here are supplied by Fosh's own Mallorcan farm.
www.marcfosh.com; tel 971 720 114; Carrer de la Missió 7A, Palma

06

ES VERGER
Grab a pew in this rustic restaurant, situated halfway up Àlaro mountain. The delicious shoulder of local lamb – perfect with one of the island's earthy reds – is one of Mallorca's essential dining experiences.
Tel 971 182 126; Camí des Castell, Alaró

WHAT TO DO

WALK THE TRAMUNTANA
Walk off some of the food and wine along the Dry Stone Route (Ruta de Pedra Sec), a 170km (106-mile) path along the Tramuntana mountain range, skirting around sandy coves and trekking through aromatic patches of wild herbs and pine forest. The walk can be broken into eight stages and there are several refuges where you can spend the night. The Consell de Mallorca hand out maps and can make refuge reservations for you.
www.conselldemallorca.net

CYCLE THE ISLAND
Rent (or bring) a bicycle and explore the famously cycle-friendly island on two wheels. It's easy to follow signposted routes in the flatlands of Es Pla or into the mountains. You could actually do this whole wine trail by bicycle, if desired.

CELEBRATIONS

BINISSALEM WINE FESTIVAL
Just after the annual harvest in September, Binissalem's wine festival concludes with the epic Battle of the Grapes: not a wine tasting but a full-on food fight – you'd be well advised to leave your best clothes at home.

POLLENCA WINE FESTIVAL
In spring (often May), the Associació Vi Primitiu de Pollença hosts a weekend of wine tasting in the convent of Santo Domino in central Pollença. It's a good opportunity to taste wines from places that aren't usually open to the public.

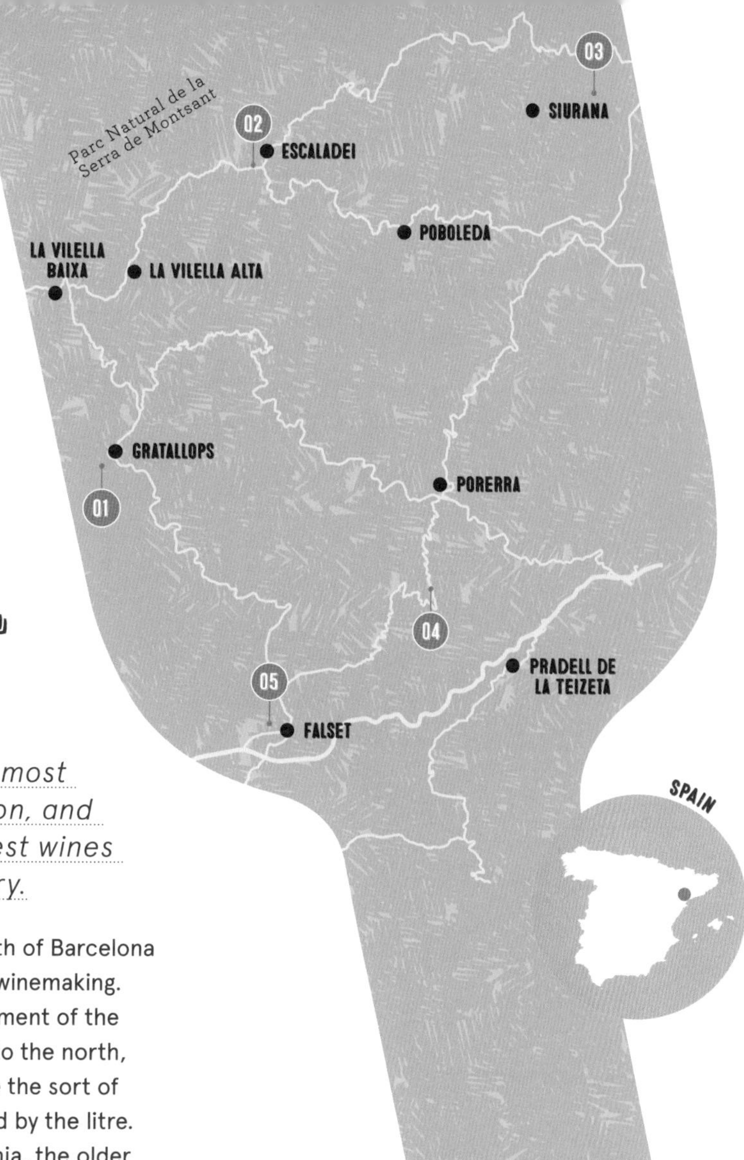

[Spain]

PRIORAT

Meet the pioneers of Spain's most adventurous viticultural region, and experience some of the wildest wines and landscapes in the country.

Less than two hours' drive south of Barcelona lies the Wild West of Spanish winemaking. Bordered by the great escarpment of the Parc Natural de la Serra de Montsant to the north, compact and rugged Priorat was once the sort of place where undrinkable wine was sold by the litre. In this impoverished corner of Catalonia, the older generation eked out a living and the young yearned to escape to Barcelona.

But in the late 1970s, a band of five long-haired pioneers, including René Barbier and Alvaro Palacios, realised Priorat's potential. On dry, sun-drenched slopes grew Carignane and Garnacha vines of 70 or 80 years old, on gradients so steep – up to 60° – that the vines delve deep into the rock to find water. The resulting yields are so low that seven to ten plants make just one bottle of Priorat. These visionary winemakers knew that hard-working vines meant complex, potent wines were waiting to be unleashed. But first they had to tame the alcohol content, which could hit 18%, and tame the land. They succeeded at the first challenge, but Priorat remains a wild and densely beautiful landscape dotted with old villages and tiny vineyards, most at an altitude of 350–400m (1200ft) but some at 900, even 1000m (3280ft).

Today, Priorat boasts one of Spain's most expensive wines, Alvaro Palacios' L'Ermita, and is a fascinating place to tour. 'It was a wine region about to disappear,' says René Barbier, 'but now young people are coming back and they're proud to be part of it.'

GET THERE
The closest airport is Barcelona's; it's an easy 2hr drive south to Priorat via Tarragona.

01 Clos Mogador vineyards

02 Escaladei Carthusian monastery

03 Catalonian *porrones* (wine pitchers)

04 Priorat countryside

01 CLOS MOGADOR

Born in the port city of Tarragona, Clos Mogador's shaggy-bearded founder René Barbier would go hiking in Priorat as a youth: 'It was pure, untouched and wild.' Inspired by the hippie movement of the late 1960s and 1970s he returned, as he says, 'to plant the seeds of a better world'. The place he chose to settle with his wife Isabelle was just outside the hilltop village of Gratallops. More a wine safari than a wine tour, Clos Mogador will drive you to two vineyards before a wine tasting. An exploration of its steeply sloping vineyards is a geology lesson: the ground high up here consists of shards of schist, a rock known locally as *llicorella*. Gnarled old Carignane and Garnacha vines emerge from this poor soil, which lends Priorat wines their unique mineral edge.

The next lesson is botany, for René Barbier is obsessed with biodiveristy. He wants wine to be part of the natural environment, which is why 30 types of flowers and wild herbs such as fennel, rosemary and thyme flourish in his vineyards. Almond, olive, fig, cherry and walnut trees punctuate the vines. 'We want nature to be free,' he says (up to a point: wild boar from the forest are kept away from the grapes by electric fences).

His self-professed hippie philosophy extends to his wine-making, now aided by his son René Jnr and his brother Christian. They use only rainwater for irrigation and plant wheat around the wines, which, when cut, reduces evaporation. Grapes are picked and sorted by hand in September, then pressed using a small cast-iron press, allowing the winemaker to keep tasting the must; just 50,000 bottles are produced in an average year. And although Barbier is one of the forefathers of Priorat, and has studied its vineyards and villages for dozens of years, he's still experimenting.

So, how best to enjoy Priorat's big personality? It's a wine that benefits from being decanted; these are not light wines and, once allowed to breathe, a classic Priorat blend

of Garnacha, Carignane with a dash of Shiraz and Cabernet Sauvignon, such as Clos Mogador's eponymous wine, will reveal a smoky swirl of cherries, cedar and herbs. 'Wine is all about passion and patience,' says René. 'What I want is for what you have in your glass to be what you see outside.'
www.closmogador.com; tel 977 839 171; Camí Manyetes, Gratallops; Mon, Thu, Fri & Sat by appointment

02 CELLERS DE SCALA DEI

Continuing northeast from la Vilella Baixa to the quiet town of Escaladei, you'll reach one of the oldest wineries in Priorat. Monks introduced vines to the region in the 12th century and at this monastery, founded in 1194, you can tour the cellars in which wines are still aged. The monks seem to have taken the responsibility of winemaking very seriously and by 1629 they had produced a manual noting the varieties that grew best (Garnacha and Mataró, also known as Mourvèdre). Places to eat here include El Rebost de la Cartoixa.
www.cellersdescaladei.com; tel 977 827 173; Rambla Cartoixa, Escaladei; daily, tour timings vary

03 MARCO ABELLA

The drive to Porrera is spectacular, crossing a high plateau from where you can watch the weather rolling across the hills. The altitude of Priorat's vineyards is as important a factor as the soil and the sun, and few vineyards are loftier than those of Marco Abella, a winery owned by a family with roots in the region going back to the 15th century. The premium tour takes in La Mallola vineyard at 680m (2230ft) above sea level, where Carignane and Garnacha grapes grow, cooled by the Llevant wind from the Mediterranean. Bottles of Marco Abella's highly regarded wines bear artwork by Barcelona-born abstract artist Josep Guinovar.
www.marcoabella.com; tel 977 262 825; Carretera Porrera a Cornudella del Montsant, Porrera; tours Wed–Sun by appointment (other times by arrangement)

05 Tending the Clos Mogador vines

06 Cellers de Scala Dei

04 FERRER BOBET

Best friends Sergi Ferrer-Salat and Raül Bobet began making wines together at the turn of the millennium and with that auspicious date came a desire to do things a little differently. 'We wanted to emphasise freshness and elegance above all, we wanted to create a more "Burgundian" style of Priorat, more, let's say, drinkable,' explains Sergi, 'but without losing the incredible complexity and minerality of the century-old Carignane vines.' To achieve this, they found a base high up in Porrera, one of the coolest corners of Priorat, and planted away from the sun's fierce glare. 'By avoiding over-ripeness, our wines better express both the terroir and the minerality of Priorat,' say Sergi. 'Also, the winery is gravity-fed, which makes for a much more gentle treatment of the grapes.'

Ferrer Bobet's contemporary winery – rare in these parts – follows the contours of the hilltop, like the vines below. Designed by Catalan architects Espinet-Ubach, it's highly energy-efficient, with glass walls providing mesmerising views of the surrounding hills from the visitor area, while down on the subterranean winemaking levels the alchemy happens, with gravity doing a lot of the work.

Sergi and Raül remain ambitious: 'Despite the prestige of Priorat's wines and their international success, the region has to work harder on attracting wine-lovers, while preserving the beauty of this landscape and its quiet lifestyle.'

www.ferrerbobet.com; tel 609 945 532; Carretera Falset a Porrera km6.5; Mon–Fri plus first & third Sat by appointment $

05 EL CELLER COOPERATIVA DE FALSET

Back in Falset, stop by the town's wine cooperative, built in 1919 by architect Cèsar Martinell, a disciple of Gaudì, though eschewing his flourishes for a cathedral-inspired modernism.

A guided tour at midday daily except Wednesdays (March to December) uses an actor to usher visitors through a history of winemaking in the region. In summer you can taste the cooperative's wines – made by Marta Ferré – in the main foyer, in front of the great wooden tanks used for vermouth, another of the cooperative's products.

www.etim.cat; tel 977 830 105; Carrer Miquel Barceló 31, Falset; daily except Wed by appointment, closed Mon & Jan–Mar $

ESSENTIAL INFORMATION

WHERE TO STAY

CLOS FIGUERAS
One day in 1997, René Barbier introduced Christopher and Charlotte Cannan to an abandoned vineyard north of Gratallops – it is now the esteemed Clos Figueras winery. In addition to tours, tastings and meals, Christopher and Charlotte offer several B&B rooms.
www.closfigueras.info; tel 977 262 373; Carrer de la Font 38, Gratallops

CAL COMPTE
Plumb in the centre of the region, in the village of Torroja, this guesthouse is allied to the Terroir Al Límit winery. With bedrooms in a baroque 16th- to 18th-century building and open-air dining, this is a romantic spot to stay in Priorat.
www.calcompte.com; Carrer Major 4, Torroja del Priorat

CAL LLOP
Cosy Cal Llop in the hilltop village of Gratallops has balconies overlooking Priorat's vine terraces.
www.cal-llop.com; tel 977 83 95 02; Carrer de Dalt 21, Gratallops

WHERE TO EAT

RESTAURANT LA COOPERATIVA
Seasonal, locally produced food (the olive oil is from Falset) and Priorat wines without a markup ensure that La Cooperativa is a popular place to eat in Porrera. This is rural Spain – expect rabbit, lamb or even boar.
www.restaurantla cooperativa.com; tel 977 82 83 78; Carrer Unió 7, Porrera

EL CELLER DE L'ASPIC
Chef Toni Bru cooks updated Catalan classics at this Falset fixture. The wine list is notably good.
www.cellerdelaspic.com; tel 977 831 246; C/Miquel Barceló 31, Falset

WHAT TO DO

Pack your hiking boots, because Priorat is laced with outstanding trails, especially to the north, along the edge of the Parc Natural de la Serra de Montsant. From an eyrie at the top of this wall of rock you look out over the whole of Priorat's amphitheatre; a network of via ferrata cables aids novice climbers. Book guides and get maps from the visitor centre in La Morrera de Montsant. It's not only climbers who head to the cliffs in the northeast of the region around Siurana – this lost-in-time clifftop village is a fabulous place in which to explore and enjoy the views from the restaurant.

CELEBRATIONS

In early May the annual wine fair takes over Falset for a week, with more than 60 producers showcasing their wines, including Clos Mogador and Alvaro Palacios. It's a hugely sociable event, with a programme of events plus organised activities for children. A *Bus de Vi* (wine bus) runs from Tarragona and Reus to Falset.
www.firadelvi.org

[Spain]

RÍAS BAIXAS

For sea, shellfish and a very special wine, make a pilgrimage to this rugged nugget of Galicia.

Perched just above Portugal, on the Iberian peninsula's northwest coast, Galicia is a corner of Spain with a rich seafaring tradition and sought-after seafood. All year round its granite headlands are blasted by the Atlantic's wind and waves, but in between those rock bulwarks lie sheltered inlets – or rías. It is these that lend their name to Galicia's Rías Baixas wine region, the most interesting of the five *Denominación de Origen* (DO) regions in Galicia.

They do things differently in Galicia. Here, vines are suspended 2m (6.5ft) above the ground from granite pillars, all the better to gain extra ventilation in this humid region and prevent mould. And the grapes that hang from these canopies are also a little bit special. Stony yet fruity, the Albariño grape that grows in Rías Baixas produces a white wine that is a mesmerising alternative to Chablis (Chardonnay) and Sauvignon Blanc – and this is its heartland.

This grape is the key to how Rías Baixas' white wines manage to meld the savoury, saline flavours of their surroundings with a Viognier-like fruitiness to mouthwatering effect. On a sunny afternoon, there is no better companion to a plate of seafood – oysters and octopus are the obvious options but don't fear the local speciality of *percebes* (lobster-flavoured goose barnacles, harvested from cliffs in between crashing waves). Outside Galicia, spicy Asian food is another great partner for Albariño.

Many visitors to Galicia arrive on foot, having made the pilgrimage to Santiago de Compostela on the Way of St James. Rías Baixas is a couple of hours' drive south of the Galician capital, and the revitalising coastal setting, with its wide and sheltered beaches, backed by aromatic pine and eucalypt forests in the interior, appeals to footsore pilgrims and wine-seeking sybarites alike.

GET THERE
The closest international airport is at Santiago de Compostela. It's an hour's drive south to Cambados.

01 BODEGAS DEL PALACIO DE FEFIÑANES

Hidden in the granite cloisters of a vast, austere palace in the centre of Cambados, this winery was one of the earliest producers of wine in Rías Baixas, releasing its first bottles in 1928, and the first to be recognised as *Denominación de Origen* Rías Baixas. Today there are five subregions and this tour focuses on the original source of the Albariño grape in Rías Baixas, here in the Val de Salnés.

The Palacio de Fefiñanes dates back to the 17th century and its stone is a clue to Albariño's savoury edge: the mineral-rich, granitic soil. As owner Juan Gil de Araújo puts it: 'a wine must be loyal to its origins'.

www.fefinanes.com; tel 986 542 204; Plaza de Fefiñanes, Cambados; Mon–Sat & by appointment Sun

02 MARTÍN CÓDAX

From the tasting terrace at Martín Códax, high on a hilltop behind Cambados, visitors enjoy views over the mussel and oyster beds in the bay that produce the ideal accompaniment to its white wines.

It's all about Albariño at this cooperative, where 280 members pool harvests from their small plots every September and October – unlike most Spanish regions, co-ops are common here. A great way to understand the grape is to book a tasting, in which the wines are paired with Galician cheeses. Three wines are featured: Organistrum (the label shows the cathedral in Santiago de Compostela) is the only Albariño aged in French oak and is bottled in small batches. You should detect fresh citrus and apples from this one. Lías, named for the yeasty lees on which the wine sits and which add flavours of brioche, is a rounded Albariño wine with a long finish. And the Martín Códax Albariño is a bright, intense wine with aromas of green apples, fresh herbs and citrus fruits. It's not aged – this is as honest a taste of the Val de Salnés as it gets. Three wines made from the same grape but deliciously different.

www.martincodax.com; tel 986 526 040; Burgáns 91, Vilariño; Tue–Sat by appointment, also open Mon Jul–Sep

01 Martín Códax winery

02 Pazo de Rubianes

03 Vineyard at Pazo de Rubianes

03 PAZO DE SEÑORANS

You can tell the importance of a property by the size of the *horreo* – and the *horreo* at Pazo de Señorans has 10 pillars per side. It's a big one. A *horreo* is the outdoor food store for a house, raised above the ground to stop vermin getting in. The *horreo* at Pazo de Señorans indicates that this was a wealthy estate, yet in the 1970s the building was a ruin. It was restored by Marisol Bueno and Javier Mareque and, now at the heart of the Rías Baixas DO, it celebrates its 30th anniversary in 2020. Before you try the wine, the *pazo* (country house) itself is a fascinating place to explore: pagan and Christian symbols above the gate show typically pragmatic loyalties; a small chapel is decorated with palm trees, a symbol of travelling the world. And there's also a panic room where Edward, the last king of Portugal, is thought to have hidden.

In Galicia, explains Javier Izurieta Romero, land was more important than money. 'Family inheritances split each parcel of land so none could be sold without the agreement of all. Each family had its own vines, its own pig and a cow. They optimised the land, growing potatoes beneath the vines.' The result is that Pazo de Señorans is supplied with grapes by 110 growers who work 400 parcels of land. The sugar and acidity of the grapes is measured at each. 'If you do a good job in the vineyard you can leave the wine alone,' says Javier.

The winemaking is similarly simple, with no oak used. Just three wines are made by winemaker Anna Quintella, but the difference between the years is striking. 'Albariño is not just a wine to drink young and cold,' explains Javier. As it ages, the flavours mellow from citrus and green apple (and, some years, rose petals) to nectarines and apricots, becoming more buttery. After pressing, the skins of the grapes are used to make aguardiente, a punchy digestif, in a distillery on the property. Nothing goes to waste in Galicia.

www.pazodesenorans.com; tel 986 715 373; Lugar Vilanoviña, Meis; by appointment $

04 PAZO DE RUBIANES

While there may be more than 20,000 small plots of vines and 6500 growers in Rías Baixas, this 15th-century ducal palace in the town of Rubianes is the place to explore the region's largest vineyards. The official tour leads guests through the palace's botanical gardens, which were planted in the 17th and 18th centuries and now burst with trees such as giant eucalypt, camphor

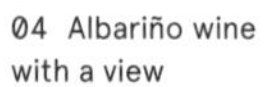

04 Albariño wine with a view

05 Pontevedra, the capital of Rías Baixas

06 *Pulpo a la gallega* (Galician octopus)

and oak, and more than 4000 varieties of camellia (the garden is a stop on the Ruta de la Camelia, see What To Do). Next, visit the vineyard, then the palace and chapel. The tour concludes with a well-earned tasting of Pazo de Rubianes' wines in the cellar, the result of agronomist Guillermo Hermo's hard work growing and selecting the best grapes.

www.pazoderubianes.com; tel 986 510 534; Rúa do Pazo 7, Rubianes; by appointment $

05 BODEGA GRANBAZÁN

To complete a loop, head back to the coast and the town of Vilanova de Arousa, just north of Cambados. When the sun's shining, the small sandy beaches here lure day trippers who then check out the tapas menus in the old town's bars. The grandest local winery is Granbazán, a short drive inland. Its Etiqueta range is the go-to Albariño, with the Etiqueta Verde being a dry, floral and decidedly salty example, and the minerality-meets-melon of the Etiqueta Ambar making it the pick of the bunch.

www.agrodebazan.com; tel 986 555 562; Lugar de Tremoedo 46, Vilanova de Arousa; tours Tue–Sat by appointment (Sun & holidays on request) $

ESSENTIAL INFORMATION

WHERE TO STAY

PARADOR DE CAMBADOS
The coastal town of Cambados is the best base for exploring the Salnés region. Set in an ancestral country house in the old quarter of Cambados, this parador has grand bedrooms (some with air-conditioning), a restaurant serving Galician specialities, and good sea views from the promenade.
www.parador.es; tel 986 542 250; Paseo Calzada, Cambados

NOVAVILA
Novavila is a rural boutique hotel from the same family as Spain's Vilanova Peña interior design brand (yes, all the furniture is available to buy). Wine is another sideline so guests are also invited to taste the family's Albariño.
www.novavilariasbaixas.com; tel 609 111 023; Santo Tomé de Nogueira, Meis

WHERE TO EAT

There are several restaurants on Rua Albergue, Rua Real and Rua Principe in Cambados. In Pontevedra, the Rías Baixas capital, many of the restaurants are concentrated in the streets of the old town.

06

EIRADO DA LEÑA
Enjoy a deliciously creative culinary experience in an intimate little stone-walled restaurant, set with white linen and fresh flowers. The Menú Curricán features ten beautifully presented courses of Galician produce for under €50.
www.eiradoeventos.com; tel 986 860 225; Praza da Leña 3, Pontevedra

BEIRAMAR
From Cambados, take a 30-minute drive to the town of O Grove on the peninsula that shelters Cambados from the full force of the Atlantic. Seafood, of course, is the town's speciality, and Beiramar is a reliable restaurant in which to enjoy scallops, oysters, fresh fish and more.
www.restaurantebeiramar.com; tel 986 731 081; Avenida Beiramar 30, O Grove

WHAT TO DO

If you're not ready for the Way of St James, the pilgrimage/hiking route across the Pyrenees that winds up in Santiago de Compostela, the Ruta de la Camelia is a gentler, more floral alternative. It follows the southern coast of Galicia, taking in 11 ornamental gardens. Two close to Pontevedra are Pazo de Quinteiro da Cruz, with its 1500 varieties of camellia, and the grand estate of Pazo de Rubianes.
www.turismo.gal

CELEBRATIONS

FIESTA DEL ALBARIÑO
On the first Sunday of August, Cambados' annual wine festival was born out of a contest between local winemakers, but is much more than that today: parades, live music by night and the naming of the year's winning wines.
www.fiestadelalbariño.com

FIESTA DEL MARISCO
In October, the region's shellfish festival is hosted by the 'seafood paradise' of O Grove. Scoff local oysters while quaffing Albariño, as the town's seafaring culture is celebrated with music, dance and sport.
www.turismogrove.es

01

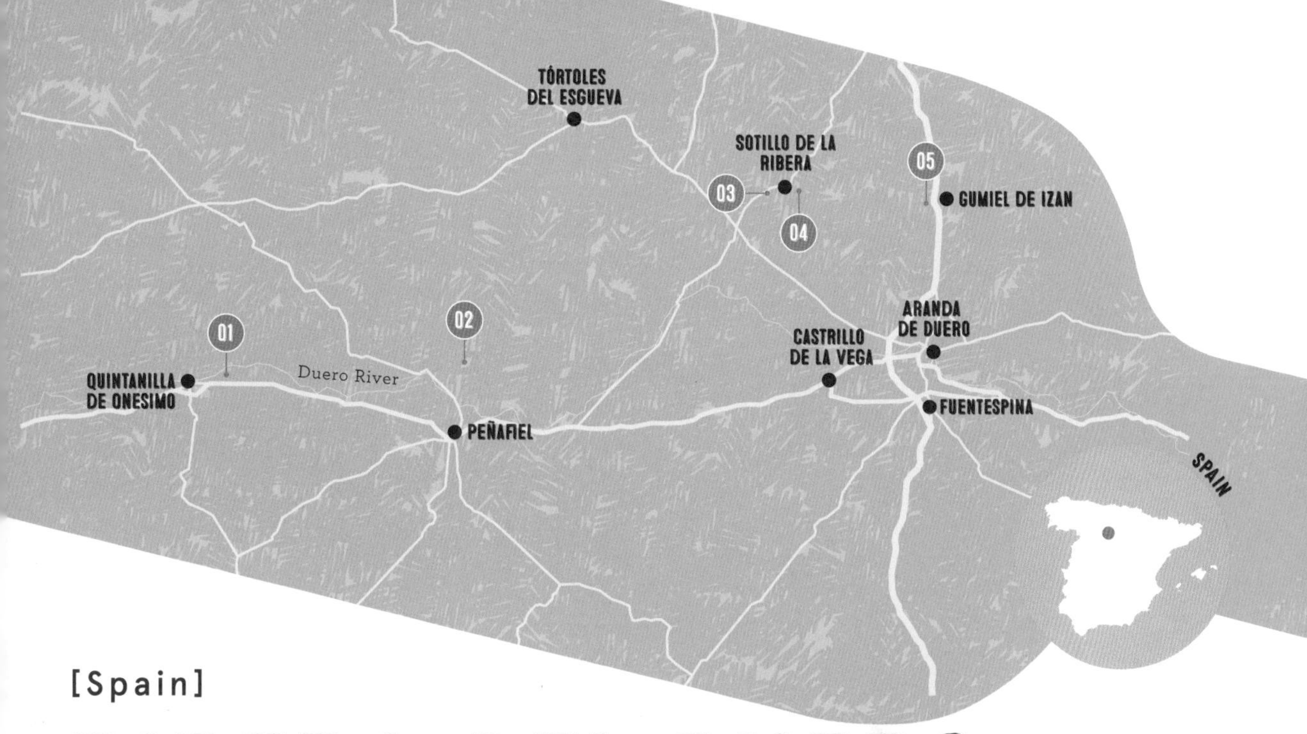

[Spain]

RIBERA DEL DUERO

Enjoy Tempranillo and slow-cooked lamb in the rustic, riverside region that threatens to usurp Rioja as the source of Spain's most sophisticated red wines.

In a country where traditions are slow to change, some of the wineries in Ribera del Duero (as they are in Priorat too) are at the frontier of winemaking. Just two hours' drive north of Madrid, what Ribera del Duero lacks in scenic majesty – these are Spain's high plains, without the sea views of Rias Baixas, the mountains of Rioja and the rugged valleys of Priorat – it makes up for with its wonderful wines. Violet when young, the Tempranillo – the grape attraction here – is seemingly inkier than Rioja, with a flavour that is less dependent on oak, more open to the winemaker's influence. It was only in 1982 that Ribera del Duero gained a *Denominación de Origen*: 'This is a young DO,' says winemaker Raphael Cherda. 'We can try new things, we're more flexible than Rioja.'

Thanks to its altitude – this is one of the highest winegrowing places in the world – the growing season in Ribera del Duero is short and it can be very hot during the day and very cold at night. But it's this variation, of up to 25°C (77°F), that creates the special conditions for producing incredible wines from vineyards along the Duero River (known as the Douro in Portugal, see p202). Those vineyards include Vega Sicilia, source of Spain's most famous wine, from plots developed by Eloy Lecanda y Chaves in 1864.

Younger Riberas taste fresh and are great with ham and fish. The Crianzas, aged for 12 months, are a perfect accompaniment to the local speciality of *lechal al horno* (slow-cooked lamb).

Make no mistake, however: aside from the wine, this is an undeveloped expanse of Castile and León. The landscape of bare plateaux, sometimes topped with a gimlet-eyed castle, is bleak; the towns and villages are functional. But the wines? The wines are wonderful.

GET THERE
Madrid is the closest city to Aranda del Duero, which makes the best base for a weekend.

❶ VIÑA MAYOR

Depending on where you're staying, it makes sense to start at the far end of Ribera's 'Golden Mile', a strip of wineries strung along the very fast and straight N-122 (be warned!) before venturing off the beaten path. Viña Mayor near Peñafiel offers an excellent introduction to the region, with a guided tour of the winery explaining such nuances as the difference between American and French oak barrels (it's all about the pores in the wood), before concluding in a glass-fronted tasting room overlooking the orange, iron-rich earth and vines stretching down to the road. Wines are produced in both a classic style and, under the Secreto label, a modern style; a three-hour tasting course with Gema García Muñoz highlights the differences. Other activities include learning how to blend a wine or pick and press grapes.

www.grupobodegaspalacio.es; tel 983 680 461; Carretera de Valladolid a Soria km325, Quintanilla de Onésimo; by appointment

❷ COMENGE

'This is nature, but controlled,' says Comenge's winemaker Raphael Cherda. The winery earned its reputation as one of Ribera del Duero's best, thanks to an approach that blends technology with ecology. An example: Comenge uses its own natural yeast – but that yeast was selected as the best for the wine at the University of Madrid (where Raphael studied) from 300 samples taken from its vineyards.

Comenge is a small, young winery, with 32 hectares (79 acres) of Tempranillo vines, half of them surrounding the modern building, which was built by the Comenge family in 1999. At 800–900m (2600–3000ft), the tasting room and terrace overlook the valley of the River Duero, and in turn are overlooked by Curiel's hilltop castle. Pesticides and herbicides are not used; instead it lets the grass grow (the competition for water is good for the vines) then cuts it to insulate the vines later in summer. Comenge is looking for fewer clusters and smaller grapes. 'We don't try to make the same wine every year,' says

01 Aranda de Duero

02 Santa María church, Aranda de Duero

03 Viña Mayor winery

04 Ribera del Duero wines

Raphael. 'It's important to express what happened in the vineyard that year and every year is different. The most important thing is the grape.'

Grapes are picked by hand in autumn; the local Tempranillo is quick to ripen in the warm, sunny days but the cold nights help the wine retain its distinctive colour. Raphael's skill and the winery's attention to detail ensures that the Don Miguel, made from grapes grown in the highest plots, strikes a great balance between fruit and toasty oak flavours; at €33 it's half the price of comparable wines.

www.comenge.com; tel 983 880 363; Camino del Castillo, Curiel de Duero; Mon–Fri by appointment, open two Saturdays per month Feb–Dec

03 FÉLIX CALLEJO

It's a family affair at Félix Callejo: father Félix is from Sotillo and returned to the village in 1989 to start his winery, which now employs both his daughters (Beatriz is on the business side, Noelia is a winemaker) and his son José, who studied winemaking at the University of Madrid with Raphael Cherda from Comenge. His mission was not only to make better wine (isn't that the goal of every winemaker?) but to revive an old-fashioned local white grape called Albillo. Previously in this part of Spain, families made their own wines and had their own cellar. They would plant a small plot with eight or nine vines of white, for eating, around a walnut tree. But the white grapes they didn't eat were mixed with red wine to make 'Clarette' – named for its clear colour – which they pressed with their feet. 'Our grandparents' generation,' explains Beatriz, 'drank a lot of wine so it had to be lighter.' Callejo has a 3-hectare (7-acre) plot of these white grapes and in 2018 made its fifth season of El Lebrero.

But Félix's main focus is on Tempranillo, which is grown on limestone at a height of 860–930m (2820–3050ft). This elevation means that it's cold at night, so the local variety of Tempranillo has developed a thicker skin, allowing more flavour to be extracted at pressing. And it's not the only advantage this part of Ribera

05 Bodegas Ismael Arroyo vineyards

06 Raphael Cherda of Comenge

del Duero enjoys over Rioja: the weather is warmer and drier. Callejo uses only its own grapes and some of the vines are now 25 years old; these go into a Gran Reserva that is aged for 24 months.

www.bodegasfelixcallejo.com; tel 947 532 312; Avenida del Cid km16, Sotillo de la Ribera; tours noon Mon–Sat, by appointment $

04 BODEGAS ISMAEL ARROYO

To understand even more of Ribera del Duero's history, continue to Bodegas Ismael Arroyo on the other side of Sotillo (the village lends its name to Arroyo's award-winning ValSotillo wine). As was usual in many of the region's villages, every building in Sotillo had a *lagar*, a cave for storing wine, tunnelled into the hills in the village. Owner Miguel Arroyo explains: 'Villagers would carry wine in goatskins, which, when full, were just the right size to lift to the *lagares* after the grapes had been pressed.'

These cellars were built between the 16th century and the 1960s and Arroyo is one of very few wineries still using its family cellar, which is in the hill behind the winery and the location for tastings. 'You can understand how important wine was culturally,' says Miguel, 'because the stones inside were expensive and only otherwise used for palaces and churches.' Inside the narrow cellar it's cool – always 11°C (52°F), winter or summer – and, when lit by flickering wall lights, slightly eerie. The family started making their own wine in 1979, breaking away from the local cooperative. They still do things the old-fashioned way: no pesticides and only the natural yeasts present in the grape skins are used.

www.valsotillo.com; tel 947 532 309; Los Lagares 71, Sotillo de la Ribera; Mon–Sat by appointment $

05 BODEGAS PORTIA

Wine meets architecture on the road into Aranda de Duero from the north. Here stands Portia, a vast, squat, high-tech winery designed by British architect Norman Foster. It's a stark contrast to wineries such as Arroyo, and is one of two star-architect-designed projects in Ribera del Duero; the other is Richard Rogers' Bodegas Protos in Peñafiel. But while Protos resembles a provincial airport, Portia has more of a spaceport look to it – which is why we prefer it. Portia's wines are good examples of the region: at the 2019 Concours Mondiale de Bruxelles, an annual international blind tasting, Portia scored a gold medal. At the same competition of 9000 wines from five continents, Spain and Ribera del Duero headed the medal table.

Portia is also rather creative when it comes to the events front: in addition to the regular tours and tastings, it offers aroma workshops and stargazing evenings.

www.bodegasportia.com; tel 947 102 700; Antigua Carretera N-1, Gumiel de Izán; tours daily Burdett; Tastings by appointment, 5pm Fri, 10.30am & 1.30pm Sat $

ESSENTIAL INFORMATION

WHERE TO STAY

For a city stay, Aranda de Duero makes a good base, with the widest selection of accommodation. Some of the bodegas have also started to offer accommodation but options are limited in the countryside.

HOTEL TORREMILANOS
West of Aranda de Duero, this hotel is based in the Bodegas Peñalba Lopez. Rooms are spacious, the location is good and you can tour the bodega.
www.torremilanos.es; tel 947 512 852; Finca Torremilanos, Aranda de Duero

WHERE TO EAT

EL LAGAR DE ISILLA
In central Aranda de Duero, El Lagar de Isilla is the go-to place for traditional, ribsticking Ribera cooking – slow-cooked lamb, not many vegetables – which is the perfect companion to the region's wines. The restaurant is part of a winery business that also owns ancient cellars beneath Aranda de Duero. These are open to visitors.
www.lagarisilla.es; tel 947 510 683; C/Isilla 18, Aranda de Duero

MOLINO DE PALACIOS
If it's autumn, fans of fungi should head for this converted watermill in the west of the region. Wild mushrooms are a seasonal obsession in northern Spain and Molino de Palacios does them best.
www.molinodepalacios.com; tel 983 880 505; Avenida de la Constitucion, 16, Peñafiel

WHAT TO DO

As there's not much happening on the *meseta* (plateau), most locals go to Burgos, which might seem like a typically sombre northern-central Spanish city but has plenty of good restaurants and nightlife along C/de San Juan, C/de la Puebla and C/del Huerto del Rey, northeast of the city's standout Gothic cathedral.

CELEBRATIONS

Sonorama is an annual music festival hosted in Aranda de Duero each August – it must be one of the few pop, rock and dance music festivals in the world to also feature wine-tasting courses and events in the city's underground wine cellers.
www.sonorama-aranda.com

06

01

[Spain]

RIOJA

Hemmed by mountains, sustained by adventurous Basque cuisine and home to spectacular wineries, Rioja is arguably the world's most rewarding wine region.

Rioja is Spain's rock-star region, the Jagger to Ribera's Richards. It's moneyed, flamboyant, and fantastic fun for a wild weekend away.

Firstly, it was blessed with natural good looks: the region sits at the foothills of the Cantabrian mountains, beyond which lies the Basque country and the city of San Sebastián. The mountains are a barrier to clouds from the north, creating a sunny microclimate. Next, it had perfect timing, hitting the heights as French wine faltered, and attracting wealthy investors who splashed out on star architects (more on this later), state-of-the-art wineries and winemaking talent. The result is that there are now more than 500 wineries in Rioja. Not all are open for tastings, but those that do offer some of the most fascinating visitor experiences in the wine world.

GET THERE
Bilbao's international airport is 1hr 30min north of Logroño by car.

The River Ebro flows eastward through the region, on its way to the Mediterranean, passing through Logroño. This university city is the fulcrum for Rioja's three subregions. To the south is Rioja Baja, north of the river is Rioja Alavesa, which includes the fortified hilltop town of Laguardia, and to the west of Logroño is Rioja Alta; the focus of this Wine Trail falls on the latter two regions.

Laguardia, just north of Logroño, makes a good base for exploration. From here it's easy to reach Haro, where Rioja's original winemakers set up shop in the 1800s. Tradition is still at the heart of Rioja's wine but the old-fashioned leathery Tempranillo is fading away in the face of fresh competition from Ribera del Duero. Its wineries range from fascinating time warps to engineering marvels and contemporary curios.

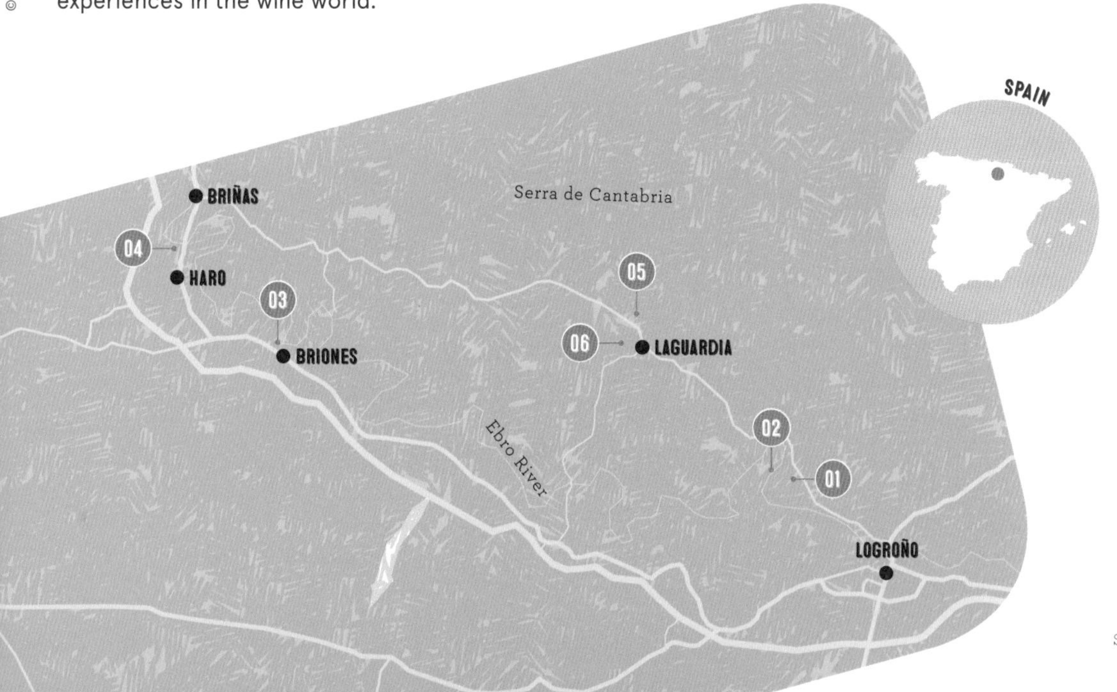

01 Zaha Hadid annexe at López de Heredia Viña Tondonia

02 Casa Primicia vineyards

03 Horse-drawn plough at Bodegas Ruiz de Viñaspre

04 Vivanco winery

01 VIÑA REAL

Seven years in the making, when Viña Real was completed in 2004 it was one of Rioja's first modern wineries. And it's an engineering marvel on a grand scale, courtesy of Bordeaux architect Philippe Mazières, whose father was a winemaker. First, a corner was cut out of a tabletop mountain between Laguardia and Logroño. Two tunnels were bored 120m (390ft) deep into the remaining mountain using the machines that excavated the tunnels of Bilbao's metro system. Then a 56m-wide (184ft) barrel-shaped building was sunk into the levelled-off corner. In the centre of this twin-storey circular room is a revolving crane arm that moves huge vats around, using gravity to pour grape juice from one to another. Head winemaker María Larrea monitors everything from a laboratory reached by a walkway. Think winery-meets-Bond-villain-lair.

Beneath the production area is a raw concrete bunker where 2000 barrels of the Reserva are stored. In the cave there are 14,000 barrels of the Crianza, which is aged for at least two years, one of which must be in a barrel. Viña Real is part of the CVNE group, which started in 1879 when French winemakers relocated to Rioja, bringing with them barrel-ageing techniques. *www.visitascvne.com; tel 941 304 809; Carretera Logroño-Laguardia km4.8, Araba; daily by appointment*

02 CONTINO

Cradled in a loop of the River Ebro, Contino is a château-style, single-estate vineyard, part of the CVNE (Compañía Vinícola Norte Espana – the Cune on the bottle is a misprint) empire. But it's a very different experience to Viña Real on the hill nearby. The stone property, just outside Laserna, is sheltered – ideal for sitting outside in the shade on old millstones, listening to the birdsong with a glass of the white Rioja to catch your breath. Contino's vineyards reach all the way down to the river Ebro, past ancient olive trees (one is 800 years old).

Seven grape varieties are permitted in Rioja wine, four red and three white: Tempranillo,

Garnacha (or Grenache), which adds a bit of weight, Carignane and Graciano are the red grapes. Contino is notable for having developed the rare and delicate Graciano grape and produces a 100% Graciano wine, which is worth comparing with a typical Tempranillo-driven Rioja.
www.visitascvne.com; tel 941 304 809; Finca San Rafael, Laserna; Mon–Sat by appointment

03 VIVANCO

Not satisfied with just building a modern winery on the outskirts of Briones, in the west of Rioja Alta, the Vivanco family added a restaurant and a museum (full name: Vivanco Museum of the Culture of Wine). And it's not a half-hearted effort: with 4000 sq metres (43,055 sq ft) of space and items from the family's personal collection spanning 8000 years of winemaking, from amphorae to art by Joan Miró, you're guaranteed to learn something about human ingenuity (though you might want to skip a few of the 3000 corkscrews). At weekends tasting courses explore Vicanco's wine. The winery itself is next to the museum and underground. 'I've always felt that our wines had to tell a story,' says winemaker Rafael Vivanco, but the museum also does a great job of telling it.
www.vivancowineculture.com; tel 941 322 323; Carretera Nacional N-232 km442, Briones; see website for opening times & tours

04 LÓPEZ DE HEREDIA VIÑA TONDONIA

The story of López de Heredia Viña Tondonia is the story of Rioja wine. Founder Rafael López de Heredia was a Basque who lived in Chile but returned to fight for the Spanish king, lost, and was exiled to France. There he started to work for a Bayonne wine merchant. Here he absorbed tips and techniques from French winemakers. The disaster of phylloxera, a vine disease that destroyed the French industry, was a blessing for Rioja. People like Raphael returned to Spain bringing new ideas and French winemakers. Recognising the region's similarities with Bordeaux, he settled in Haro in 1877 close to the train station, invested all his money in five vats and began making table wine, selling it fast and cheaply.

Fast-forward four generations and one thing has changed at López de Heredia, now managed by María José López de Heredia: the wine is outstanding (but not as cheap). Other things haven't changed: it still uses a bunch of dried Viura vine stalks as a filter, as was traditional; it doesn't use steel vats, only 100-year-old oak vats; and it still makes (some) of its own barrels.

A tour begins in a modern annexe designed by Zaha Hadid to resemble a wine decanter.

05 Central Logroño

06 *Patatas a la Riojana* (potatoes with chorizo and onion)

07 Laguardia

But the real interest lies in the old winery next door. Here, the hand-excavated wine gallery dates from 1890 and extends all the way back to the river – workers were given 4L (7 pints) of wine a day, two of which they could drink in the winery. In its darker corners the cave is coated with penicillium, a white furry mould, which helps keep the temperature constant by absorbing humidity.

The last stop is in the tasting room to sample López de Heredia's Viña Bosconia, a five-year-old Burgundy style wine, and Viña Tondonia, a six-year-old Bordeaux style wine. After more than a century of practice, both are sublime.

www.lopezdeheredia.com; tel 941 310 244; Avenida de Vizcaya 3, Haro; Mon–Sat by appointment

05 BODEGAS RUIZ DE VIÑASPRE

This family-owned winery is set in the foothills between Laguardia and the Cantabrian mountains to the north. Its neighbour, Bodegas Ysios, is one of Rioja's most iconic wineries, thanks to its spectacular wave-like roof, designed by Spanish architect Santiago Calatrava to reflect the mountain backdrop. Sadly, the roof proved to be less waterproof than desired, which kept lawyers busy. Ruiz de Viñaspre is a smaller operation, with all the Tempranillo grapes coming from the winery's own vineyards. 'You can only make the best wines from the best grapes,' says Ainhoa Ruiz de Viñaspre, who now runs the winery with her sister Jaione. 'Our vineyards are in Laguardia and Elvillar, in the heart of Rioja Alavesa, sheltered below the Sierra Cantabria. The unique microclimate makes it one of the best vine-growing areas in the world.'

The Ruiz de Viñaspre family is at the heart of the winery: 'Being from Rioja means everything. That is why it is always someone from the family who receives visitors. We want them to feel that our house is theirs.'

www.bodegaruizdevinaspre.com; tel 945 600 626; Camino de la Hoya, Laguardia; daily by appointment

06 CASA PRIMICIA

According to legend, King Sancho Abarca of Navarra once climbed a hill at the foot of the Cantabrian mountains, which overlooked the River Ebro and what is now Rioja. Recognising the hill's strategic importance, a few months later he founded La Guardia de Navarra on its top. The date was AD908. More than a thousand years later and Laguardia offers a (very popular) glimpse into the past. The hill is riddled with unexplored tunnels and blessed with beautiful architecture.

Casa Primicia, the 'first house', is the oldest property in the medieval hamlet, dating from the 15th century. It was here that grapes taxed from the local area were stored. From the 16th century, wine was made on the site and the restoration of the building, which owner Julián Madrid began in 2006, has revealed how it was done. And the tunnels that form a subterranean twin town are perfect cellars for storing Bodegas Casa Primicia's own wines. Visits include tastings and tours of the building.

www.casaprimicia.com; tel 945 600 256; C/Páganos 78, Laguardia; Wed–Mon by appointment

ESSENTIAL INFORMATION

WHERE TO STAY

If you're based in the walled hilltop town of Laguardia you'll be able to reach all the featured wineries easily. The town is popular with tourists but it's easy to escape to Logroño for better-value meals. There are several hotels in Laguardia, if you have the budget.

CASA RURAL ERLETXE
This guesthouse, in the thick walls of the town, is hosted by María Arrate Aguirre. She makes delicious homemade breakfasts that include honey from her bees stationed up in the Cantabrian hills.
www.erletxe.com; tel 945 621 015; Rua Mayor de Peralta 24–26, Laguardia

MARQUÉS DE RISCAL
For a less understated stay, try the Marques de Riscal hotel. The Frank Gehry–designed property, a cascade of ribbons of steel, lies between Logroño and Haro.
www.hotel-marquesderiscal.com; tel 945 180 880; C/Torrea 1, Elciego

WHERE TO EAT

LA TAVINA
This wine club in Logroño serves 60 wines by the glass (at shop prices) and modern tapas plates. In winter it hosts group tastings and courses.
www.latavina.com; tel 941 102 300; C/Laurel 2, Logroño

LA COCINA DE RAMON
Chef Ramón Piñeiro updates traditional Spanish recipes at his restaurant in central Logroño. There's a three-course set menu for €25 or his 'Walk through the Market' menu for €60, which is composed of creative dishes of market-fresh ingredients. And the impressive wine list features 50 labels from Rioja.
www.lacocinaderamon.es; tel 941 289 808; C/Portales 30, Logroño

WHAT TO DO

SAN SEBASTIÁN
One of Europe's most delightful coastal cities is the epicentre of cutting-edge gastronomy in Spain. Well worth including on a food and wine pilgrimage, it has hundreds of bars serving pintxos, the Basque equivalent of tapas, and a constellation of garlanded restaurants, including Arzak.

IZKI NATURAL PARK
Pack a pair of hiking boots (or cycling shoes) and head into the Izki Natural Park, north of Laguardia, on one of the many signposted hiking trails that weave through the Pyrenean oak forests.

CELEBRATIONS

If you're in the area in October, the annual mushroom festival at Ezcaray is a treat for fungi fans, with autumn's bounty on display, expert talks, guided mushroom-picking walks in the Cantabrian mountains and lots of tastings in the village square. More famous is the annual Batalla del Vino (wine fight – really) in Haro at the end of June. The fiesta begins on the night of 28 June with a street party then the battle commences the following morning. And in September the Rioja Wine Harvest Festival brings costumes, music and grape-crushing to the streets of Logroño.
www.batalladelvino.com

INDEX

INDEX

CONTRIBUTORS

Andy Symington has contributed to over 100 Lonely Planet titles. When he's not travelling, he enjoys sampling the wines of Northern Spain. Follow him @andysymington

John Brunton is based in Paris and Venice, contributes regularly to The Guardian and writes on his wine travels at www.thewinetattoo.com. Follow him @thewinetattoo

Regis St Louis has written over 80 guidebooks for Lonely Planet and has sipped his way through Piedmont, Rioja and many other fabled wine regions. Follow him @regisstlouis

Anne Krebiehl, Master of Wine, is a London-based freelance wine writer, judge, and author of *The Wines of Germany*. Follow her @AnneinVino

Kerry Walker is an award-winning travel writer specialising in Central and Southern Europe. Follow her @kerrychristiani and www.its-a-small-world.com

Robin Barton is Lonely Planet's Associate Publisher and for this publication has trailed the wineries of England, Spain, France and Sicily.

Benjamin Kemper lives in Madrid where he writes about the places that make him hungriest. Find him on Instagram @benjaminkemper and www.benjamin-kemper.com

Luke Waterson is a Wales-based fiction and travel writer specialising in the UK, Eastern Europe and Latin America. Find him on www.lukeandhiswords.com

Sarah Ahmed is a leading authority on and champion of Portuguese wines; in addition to writing for respected titles, she publishes www.thewinedetective.co.uk

Caroline Gilby, Master of Wine, specialises in Eastern European wines and is author of *The Wines of Bulgaria, Romania and Moldova*. Follow her @carolinegilby

Peter Dragicevich is a Lonely Planet travel writer, based in Auckland, with family connections to the wine industry in both New Zealand and Croatia.

Tara Q. Thomas is a Brooklyn-based food and wine writer and executive editor at Wine & Spirits magazine. Follow her culinary musings @TQThomas

First Edition
Published in May 2020 by Lonely Planet Global Limited
CRN 54153
www.lonelyplanet.com
ISBN 978 1 78868 946 5

Printed in China
10 9 8 7 6 5 4 3 2 1

Managing Director, Publishing Piers Pickard
Associate Publisher Robin Barton
Commissioning Editor Dora Ball
Art Direction Daniel Di Paolo
Layout Tina García
Editors Dora Ball, Monica Woods
Cartographer Wayne Murphy
Image Research Regina Wolek
Proofreading Karyn Noble
Print Production Nigel Longuet
Cover Image: Val d'Orcia, Italy. Image © Francesco Riccardo Iacomino / Getty Images

Thanks to Nick Mee, Polly Thomas

Written by: Andy Symington (El Bierzo), Anne Krebiehl MW (Germany), Benjamin Kemper (Georgia), Caroline Gilby MW (Romania), John Brunton (Belgium & Luxembourg, France except Bordeaux, Italy except Liguria and Sicily, Slovenia), Kerry Walker (Austria), Luke Waterson (Hungary), Peter Dragicevich (Croatia), Regis St Louis (Liguria), Robin Barton (England, Bordeaux, Sicily, Spain except El Bierzo), Sarah Ahmed (Portugal), Tara Q. Thomas (Greece).

Lonely Planet offices
STAY IN TOUCH lonelyplanet.com/contact

AUSTRALIA The Malt Store, Level 3, 551 Swanston St, Carlton, Victoria 3053 T: 03 8379 8000

IRELAND Digital Depot, Roe Lane, The Digital Hub, Dublin 8, D08 TCV4

USA Suite 208, 155 Filbert St, Oakland, CA 94607 T: 510 250 6400

UK 240 Blackfriars Rd, London SE1 8NW T: 020 3771 5100